Effective Building Mixtures for 3D Printing

Leonid Dvorkin
Full Professor & Head
Department of Building Products Technology and Material Science
National University of Water and Environmental Engineering
Rivine, Ukraine

Marchuk V.
Associate Professor
Department of Construction Products Technology and Materials Science
National University of Water and Environmental Engineering
Rivine, Ukraine

CRC Press
Taylor & Francis Group
Boca Raton London New York

CRC Press is an imprint of the
Taylor & Francis Group, an **informa** business
A SCIENCE PUBLISHERS BOOK

First edition published 2026
by CRC Press
2385 NW Executive Center Drive, Suite 320, Boca Raton FL 33431

and by CRC Press
4 Park Square, Milton Park, Abingdon, Oxon, OX14 4RN

CRC Press is an imprint of Taylor & Francis Group, LLC

Library of Congress Cataloging-in-Publication Data (applied for)

ISBN: 978-1-041-11784-1 (hbk)
ISBN: 978-1-041-16684-9 (pbk)
ISBN: 978-1-003-68588-3 (ebk)

DOI: 10.1201/9781003685883

Typeset in Times New Roman
by Prime Publishing Services

Preface

3D printing technology is a modern, rapidly developing technology that allows to significantly reduce labor and energy costs and significantly increase the speed of construction of buildings. Effective application of this technology in construction and increasing the scale of its use along with the improvement of the necessary mechanical equipment requires the development of mixtures that meet the specific requirements of 3D printing and at the same time are characterized by the availability of raw materials, resource saving, and high performance properties.

This monograph was prepared on the basis of the results of research carried out by the authors at the Department of Construction Products Technology and Materials Science of the National University of Water Environmental Engineering (Ukraine).

The studies focus on the experimental substantiation of the properties of building mixtures and concrete based on them for 3D printing using ash and slag raw materials and other dispersed industrial wastes and also modifying additives. As a rule, experimental substantiations are carried out using mathematical modeling. The presented experimental and statistical models allow us to quantitatively assess the influence of technological factors, the effects of their interaction, and, as a result, predict the quality indicators of mortars and concrete with specified properties, and choose the optimal conditions for their providing.

The monograph is intended for construction workers and manufacturers who use or implement technology for the production of building structures and the construction of building structures using 3D printers.

It can also be useful in developing new effective compositions of building mixtures intended for 3D printing.

Preface

Contents

1

Construction Mixtures for 3D Printing

1.1 Analysis of Existing Technological Solutions for the Application of 3D Printing in Construction

3D printing is an additive technology that is successfully implemented in modern construction. Additive technologies (AM–Additive Manufacturing) are technologies for layer-by-layer creation (building up) of products based on digital 3D models.

For the first time, the "3D printing" technology was used in 1984 by Charles Hull's company to reproduce real objects using digital data.

Construction printing technology is actively developing in countries such as China, the USA, and the UK.

3D concrete printing allows [1–3];

- to minimize the duration of the technological process of manufacturing structures compared to traditional methods;
- to reduce the costs provided for by the project by minimizing waste and the use of labor;
- to provide flexibility and create three-dimensional forms that cannot be implemented using conventional technologies.

Currently, there are two ways to organize construction using 3D printers:

1. location of the printer at the construction site and construction of structures by layer-by-layer application of concrete mix in accordance with the project;
2. printing three-dimensional elements in the factory and transporting them to the site where the structures are assembled using traditional technologies.

DOI: 10.1201/9781003685883-1

By design, existing 3D printers can be divided into the followin types:

- gantry;
- bridge;
- manipulator.

Gantry-type 3D printers consist of a gantry with an extruder or nozzle mounted at the top. The concrete mixture is pushed out of the extruder by a screw, and from the nozzle by pressure created by a concrete pump (Fig. 1.1).

In bridge-type 3D printers, the extruder or nozzle is placed on a crane beam that moves along the shop floor (Fig. 1.2).

In manipulator-type printers, the extruder or nozzle is placed at the end of its boom (Fig. 1.3).

Fig. 1.1 3D - gantry type printers

Fig. 1.2 3D-bridge type printers

3D printers are equipped with electronic equipment and software.

At the current stage of development of 3D technologies, printers with a working volume of several tens of cubic meters have been developed with the technical ability to print with concrete mix.

Working mixtures for a 3D printer must meet the relevant requirements for printing the structure and future operation of a building or structure.

Currently, a wide range of materials is used for additive technologies: various polymers and rubber, steel powders, alloys of titanium, nickel, aluminum, copper, as well as ceramic and other composites.

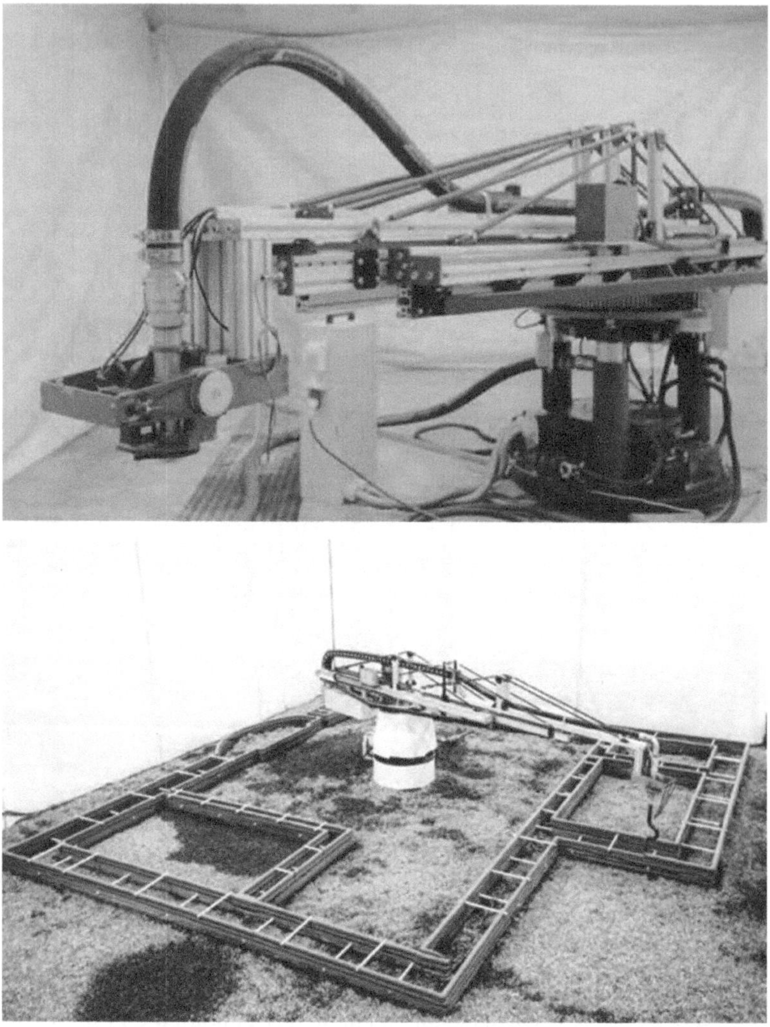

Fig. 1.3 Manipulator-type 3D printers.

For 3D printing in the construction of buildings, the most promising are quick-hardening concrete mixtures, which can contain various additives to improve the physical and mechanical properties of structural elements for various purposes [4–9].

Quick-hardening mixtures involve:

- the use of concrete mixtures with low W/C values;
- the use of additives – hardening accelerators, alumina cement, etc.;

- the addition of gypsum or the use of complex special additives;
- activation of cement mortar.

Mixtures containing steel or polymer microfiber are proposed for 3D printing [10]. In this case, it is possible to use various reinforcement schemes, including complex.

Reinforcement using bar reinforcement can be carried out both in the horizontal and vertical directions. Vertical reinforcement is installed in the technological voids of the walls after printing, followed by laying with concrete of the desired grade. Horizontal reinforcement is provided between the layers of the structure during the printing process using steel bars or flat reinforcing frames.

One of the main tasks in the implementation of 3D printing is the selection of compositions of composite building materials that meet the required operational characteristics of the building being constructed and meet the requirements of extrusion technology [11, 12]. Most often, fine-grained concrete mixtures are used for this purpose, which have the necessary strength, stiffness, frost resistance, increased adhesive and cohesive properties and hardening speed. When designing compositions of composite materials, it is necessary to take into account the features of the 3D printer. For example, when supplying the molding mixture over a long distance, its rheological properties and uniformity of structure during transportation and laying, the viability of the mixture must be preserved. To apply a thin and even layer of concrete mixture to a rough surface, which allows to ensure tight fit to the previous layer and adhesion between layers, special requirements are also required for the plastic strength index of the concrete mixture [13, 14].

To optimize the 3D printing process, it should be taken into account that:

- the adhesion strength between layers decreases with increasing time interval between laying subsequent layers;
- the building mixture must gain sufficient initial strength to withstand the next layer laid without deformation.

Performing these two constraints complicates the task of optimizing the printing speed. The time interval between two laid layers must be such as to ensure the necessary compressive strength of the molding mixture and adhesion between the layers in the early hardening period. This must be coordinated with the rhythm of the 3D printer extruder [15].

The ability of the applied layers to bear their own weight is directly proportional to the yield strength [5, 6]. When building a wall in layers, the first applied layer is subjected to maximum load. In order to ensure the stability of the wall structure during the construction process, the yield strength of the concrete mix must be sufficient to withstand this load. On the other hand, the mix must be sufficiently plastic for extrusion [7].

The yield strength of cement-based materials increases with time. It has been proven [6] that the change in yield strength is linear (Roussel model) [15] for the first 40 min, for longer times the increase in yield strength is better described by the model of Perrot et al. [16–18].

The various proposed concrete compositions for 3D printing do not offer universal solutions. Table 4.1 lists the most well-known concrete compositions for 3D printing, their properties, advantages and disadvantages [2].

1.2 Basic Properties and Test Methods of Mixtures for 3D Concreting

Properties of mixtures that ensure the operation of a 3D printer. Analysis of the proposed mixtures given in Table 4.1 allows us to conclude that their composition must meet certain criteria that are directly related to the printing characteristics and the design of the printer. In this case, the main task is to maintain the proper balance of all parameters. For example, the maximum strength of concrete involves reducing the W/C ratio by introducing plasticizer additives or reducing the water content. However, to ensure proper formability of concrete, it is necessary to maintain a certain consistency, which significantly depends on the water content. The mixture must have sufficient structural strength to provide shape and perception of the load from the subsequent layers of the structure. It is important for the mixture to remain sufficiently sticky to bond with the subsequent layer and not turn into a separate conglomerate. The mixture must also have a certain mobility to ensure its transportation and formation.

To achieve the required quality, the most important properties of the mixture are:

- formability (extrusibility),
- movability,
- structural strength (strength of the extruded layer after laying),
- shelf life (setting time),
- compressive strength,
- bond strength between layers.

Analysis of the experience of using 3D concreting in the manufacture of various structures and buildings allows us to formulate the average values of technical requirements for working mixtures (Table 1.2), which may vary depending on the type of printed structure and type of printing equipment.

Table 1.1 Some concrete composition of 3D printing.

S. No.	*Developer*	*Material*	*Density, kg/m³*	*Strength*, MPa*		*Advantages of Technology*	*Disadvantages of Technology*
				when bending	*during compression*		
1	WinSun (China)	Cement-sand mixture with demolition waste, fiberglass, and special additives	2000– 2200	8.2	34.5	Products of building materials processing, micro reinforcement are used	Requires large production areas and maintenance personnel, non-functional formwork, uneven vertical surface
2	AHT Spes-avia (Russia)	High-strength cement mixture, sand glass fiber concrete	2200– 2350	–	from 30	Micro reinforcement, materials variability	Presence of non-functional formwork, uneven vertical surface
3	Bet-Abram (Slovenia)	Shotcrete concrete with sand (<4 mm) and gravel (4-8 mm) aggregates	2300– 2350	–	–	Presence of anti-shrinkage additives	Non-functional formwork

Continuation of Table 1.1

Continuation of Table 1.1

N	*Developer*	*Material*	*Density, kg/m³*	*Strength*, MPa*		*Advantages of Technology*	*Disadvantages of Technology*
				when bending	*during compression*		
4	Contour Crafting Corp. (USA)	Cement mortar for formwork, concrete for construction	2250	–	45–50	Aligned vertical surface, reinforcement	Low strength between layers
5	ApisCor (Russia)	Fine-grained fiber concrete	2050	–	27	Micro reinforcement	Uneven vertical surface, height limitation 3.3 m
6	Loughborough University (Great Britain)	Cement concrete	2250–2350	12–13	100–110	High strength, structural reinforcement	Uneven vertical surface
7	CyBe Construction (Netherlands)	Cement concrete	2200	6	45	Aligned vertical surface, rapid setting and hardening of layers	Unsynchronized mixture supply and printer head movement

Continuation of Table 1.1

Continuation of Table 1.1

N	*Developer*	*Material*	*Density, kg/m³*	*Strength*, MPa*		*Advantages of Technology*	*Disadvantages of Technology*
				when bending	*during compression*		
8	Batiprint 3D (France)	Polyurethane foam formwork filled with concrete	–	–	–	Formwork performs a thermal insulation function	Uncontrolled geometric parameters of structures, uneven vertical surface that requires protection from external influences
9	DUS Architects (Netherlands)	Recycled plastic for formwork, lightweight concrete	–	–	–	Recycled raw materials are used	Duration of the formwork manufacturing process
10	MIT Media Lab (USA)	Polyurethane foam formwork filled with concrete	–	–	–	Formwork performs a thermal insulation function	Uncontrolled geometric parameters of structures, uneven vertical surface that requires protection from external influences

* – after testing standard molded specimens or blanks when applying a load perpendicular to the molding layer.

Table 1.2 Technical requirements for building mixtures for 3D printing.

No	*Requirement*	*Functional Purpose*	*Indicator*
1	Concrete Mixtures		
1.1	Water holding capacity, %	Ensuring the homogeneity of the mixture during manufacture and transportation	no less 97
1.2	Delamination, %		no more 10
1.3	Viscosity, Pa·s	Ensuring the possibility of transportation through pipes and uniform laying of the mixture	109–1012
1.4	Structural strength, Pa	Ensuring the necessary strength of the mixture, sufficient for laying the next layer	no less 800
1.5	Initial of setting time, min	Ensuring the possibility of laying the next layer	at least 20 after mixing
1.6	Mixture workability, by cone immersion, cm	Ensuring the ability of the mixture to be extruded and have a stable geometry	8–12
1.7	Mixture workability, by Suttard viscometer, mm		150–170
2	Properties of concrete		
2.1	Compressive strength, MPa	Ensuring the strength of structures	no less 20
2.2	Average density, kg/m^3	Ensuring thermal insulation and other physical and mechanical properties	600–2000
2.3	Adhesion, MPa	Ensuring adhesion between the layers of the mixture for strength and uniformity of the structure	no less 2
2.4	Thermal conductivity, W/(m K)	Thermal insulation of building elements	no more 0.7
2.5	Frost resistance, cycles	Ensuring the durability of structures and resistance to seasonal temperature changes	no less 100

Methods of testing mixtures for 3D concreting. The selection of the optimal composition and type of mixture requires a number of experimental studies, taking into account the necessary parameters that need *to be achieved.* At the first stage of designing concrete compositions for 3D concreting, preliminary trial tests of the mixtures are carried out, which are aimed at achieving their required formability (extrusibility), which is a critical parameter that ensures proper printing from the nozzle. Formability (extrusibility) is the ability of concrete to be extruded from the nozzle without the appearance of cracks and delaminations along the length of the beam. The necessary formability is ensured by the workability of the mixture by Standard cone penetration: 8...10 cm.

After the mixture is obtained, which is normally formed, tests are carried out to meet other requirements. It is important to use modifier additives, the introduction of which allows adjusting and improving the physical and mechanical properties of mortar and concrete mixtures, as well as concretes and mortars based on them [19–26].

The shelf life of mixtures (setting period) is determined by the immersion of the needle of the Vicat mobility by immersion of a standard cone, cm or spread on a Suttard viscometer, mm.

Structural strength is characterized by the ability to withstand the load that the mixture acquires during the formation of one layer. Structural strength can also be estimated by the number of applied layers in a structure or sample without failure.

The design of the device for determining structural strength, which we propose, is shown in Fig. 1.4. The value of structural strength is proposed to be determined by the maximum specific load that the mixture can withstand after 10, 25, and 40 minutes after mixing without deformation and failure (Fig. 1.5). The load is created using weights or sand, which are loaded into the container until the sample begins to deform.

The compressive strength of mortars is determined on manufactured beam specimens measuring 40 × 40 × 160 mm. For fine-grained concrete with a maximum grain size of more than 5 mm, cube specimens measuring 70.7 x 70.7 x 70.7 mm are tested. The bond strength between layers is determined by the value of the splitting tensile strength at the boundary between layers.

The laboratory printer (Fig. 1.6) consists of four main components:

- extruder, for printing the product;
- carriage, for moving the extruder in the horizontal and vertical directions;
- control panel, for controlling the process of printing the product from the mortar mixture;
- frame, for arranging the main components of the laboratory printer.

Technical indicators of the laboratory printer are given in Table 1.3.

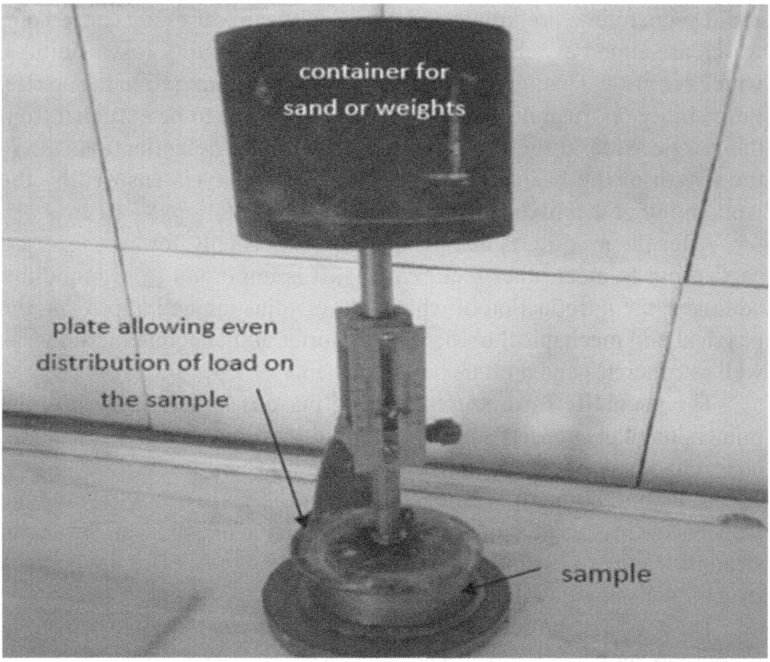

Fig. 1.4 Device for determining the structural strength.

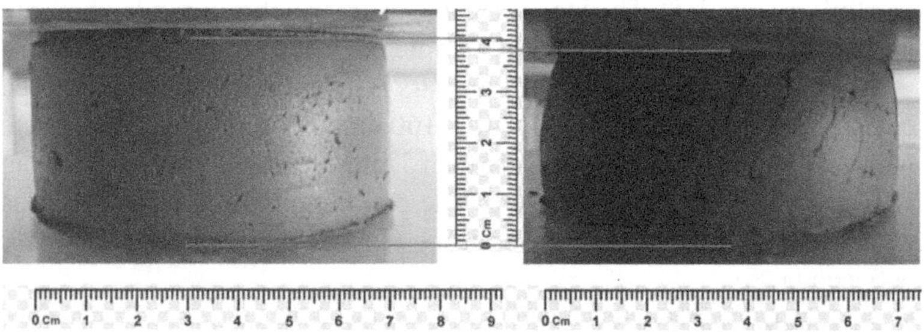

(a) (b)

Fig. 1.5 An example of determining the structural strength of extruded concrete: (a) the sample withstands the load (structural strength is provided); (b) structural strength is not provided.

Fig. 1.6 Laboratory 3D printer: 1 – electric motor of the extruder; 2 – hopper of building mixture; 3 – auger; 4 – replaceable nozzle; 5 – control panel; 6 – frequency converter of electricity; 7 – reverse motor moving the extruder in the horizontal direction; 8 – manual drive moving the extruder in the vertical direction; 9 – frame; 10 – power cable of electric motors.

Table 1.3 Technical indicators of the laboratory printer.

Main Technical Parameters of the Laboratory Printer	*Value*
hopper volume	5 l
screw rotation speed	10–30 rev/min
layer thickness during printing	10–30 mm
extruder horizontal movement speed	5 cm/s
printer productivity	30–50 l/h
extruder electric motor power	120 W
carriage reversible electric motor power	25 W
Overall dimensions	
length	1000 mm
width	500 mm
height	500 mm

2

Modified Ash-Containing Construction Mixtures for 3D Printing

2.1 Technological Properties of Cement-ash Modified Mortar Mixtures

Water consumption and water retaining capacity. The technological efficiency of mortar mixtures is largely determined by their water consumption and resistance to delamination, which is characterized by water retention capacity. Also important are the 'viability' of mixtures, i.e., the preservation of their consistency over time, the ability to entrain air during mixing and air retention during transportation and compaction.

To study the influence of the content and composition of polyfunctional modifiers (PFM), which are complex chemical additives, on the basic properties of modified cement-ash pastes, algorithmized experiments were performed according to the three-level three-factor plan B3 [22]. As PFM, compositions of superplasticizer SP-3 with air-entraining $(PFM)_1$ and water-retaining $(PFM)_2$ additives were used. The conditions for planning experiments to determine the main properties of modified cement-ash pastes are given in Table 2.1. Statistical processing of experimental data allowed us to obtain the corresponding regression equations in the coded variables, which are given in Table 2.2.

Cement-ash pastes were prepared at constant plasticity on a shaking table and the water demand of the mixtures was determined, which is necessary to achieve a cone spread of 170...180 mm. According to the standard method, the volumetric water separation coefficient (K_w, %) of the pastes was determined:

$$K_w = \frac{a-b}{a} \cdot 100\% \qquad (2.1)$$

where, a – is the initial volume of cement suspension, cm^3;

b – volume of cement suspension that has settled, cm^3.

DOI: 10.1201/9781003685883-2

Water-holding capacity of the suspension:

$$W_h^c = 100 - K_w. \quad (2.2)$$

Table 2.1 Conditions for planning an experiment when studying the normal consistency and effective viscosity of cement-ash pastes with PFM additives.

Technological Factors		*Levels of Variation*			*Variation Interval*
Natural View	*Coded View*	*– 1*	*0*	*+1*	
Pastes with additives PFM_1 (SP-3 + AeA)*					
Air-entraining admixture content, % of cement mass	$X_{1(1)}$	0	0,02	0,04	0,02
Content of superplasticizer SP-3, % of cement mass	X_2	0	0,35	0, 7	0,35
Ash-cement relationship	X_3	0	0,35	0,7	0,35
Pastes with additives PFM_2 (SP-3 + WrA)**					
Ash-cement relationship	X_3	0	0,35	0, 7	0,35
Content of water-retaining additive (cellulose ether – CE), % of cement mass	$X_{1(2)}$	0,1	0,25	0, 4	0,15
Content of superplasticizer SP-3, % of cement mass	X_2	0	0,35	0,7	0,35

Note: * AeA – air entraining additive;
** WrA – water-retaining additive

Table 2.2 Experimental statistical models of the properties of modified cement-ash pastes

Properties	*Regression Equation*
Polyfunctional Modifier PFM_1	
Water consumption, l/m^3	$W = 265{,}7 - 3{,}19X_{1(1)} - 27{,}3X_2 - 1{,}20X_3 + 0{,}74X_{1(1)}^2 - 5{,}26X_2^2 + 2{,}24X_3^2 + 0{,}88X_{1(1)}X_2 + 2{,}13X_{1(1)}X_3 + 1{,}88X_2X_3$ (2.3)
Water holding capacity, %	$W_h^c = 3{,}39 - 0{,}47X_{1(1)} + 0{,}11X_2 - 0{,}43X_3 + 0{,}254X_{1(1)}^2 - 0{,}246X_2^2 + 0{,}45X_3^2 + 0{,}113X_2X_3$ (2.4)

Continuation Table 2.2

Continuation Table 2.2

Polyfunctional Modifier PFM_2	
Water consumption, l/m^3	$W = 271{,}7 + 3{,}8X_{1(11)} - 25{,}6X_2 - 1{,}80X_3 + 4{,}01X^2_{1(11)} - 2{,}99X^2_2 + 1{,}99X^2_3 - 0{,}50X_{1(11)}X_2 + 3{,}50X_{1(11)}X_3 + 2{,}50X_2X_3$ (2.5)
Water holding capacity, %	$W^c_h = 0{,}70 - 1{,}79X_{1(11)} + 0{,}04X_2 - 0{,}26X_3 + 1{,}703X^2_{1(11)} - 0{,}147X^2_2 + 0{,}053X^2_3 + 0{,}163^2_{1(11)}X_3 + 0{,}138X_2X_3$ (2.6)

As expected from the analysis of the obtained models, the most significant reduction in water consumption occurs when the superplasticizer SP-3 predominates in the composition (Figs. 2.1, 2.2). At the same time, the most significant effect of SP-3 is observed at its dosage of 0.7% of the cement mass (reduction in water consumption to 20…22%).

The air-entraining additive, being a weaker plasticizer than SP-3, in a composition with the latter, exhibits a significant plasticizing effect. The synergistic plasticizing effect of the superplasticizer SP-3 and the air-entraining surfactant is manifested by the interaction effect of X_1 and X_2 in the water demand model, which can be explained by the different mechanisms of their thinning action.

In the mechanism of action of the superplasticizer SP-3, as is known [23–25], the effect of electrostatic repulsion of cement particles prevails, due to the fact that adsorption layers of SP molecules increase the value of the zeta potential on the surface of cement grains. The plasticizing effect of air-entraining surfactants is mainly due to the lubricating effect of the air emulsion, which significantly increases the ability of the system to deform when external forces are applied to it. Combining the above mechanisms of plasticizing action of its components in PFM allows for the appropriate thinning effect.

The addition of cellulose ether as a thickener of cement and cement-ash pastes leads to a certain increase in their water consumption. When introducing PFM, which contains the superplasticizer SP-3 along with CE, the water consumption of the mixtures decreases, however, the 'thickening' effect of CE is felt throughout the studied range of changes in the SP-3.

A significant decrease in the workability of the cement-ash pastes when using cellulose ether and a superplasticizer of the naphthalene-formaldehyde Type SP-3 (SP NF type) is due to the emergence of hydrogen

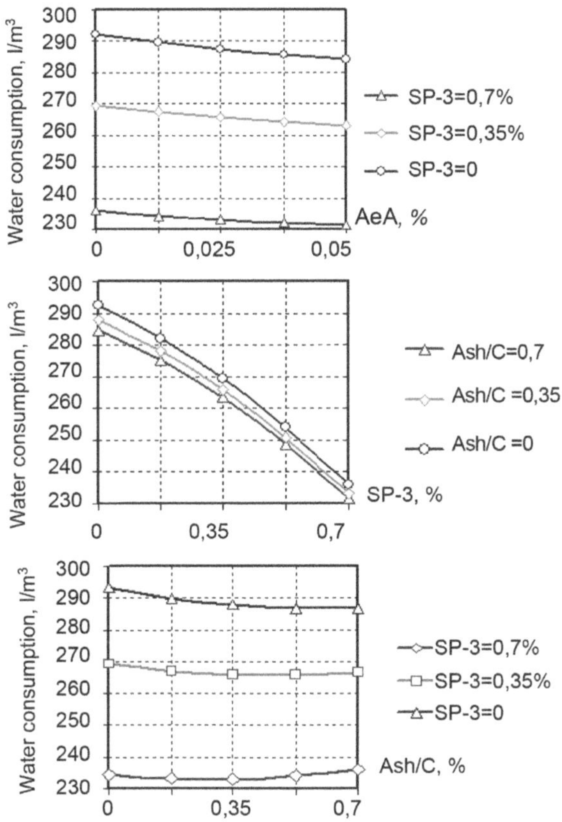

Fig. 2.1 Dependence of water demand of cement-ash mixtures on the content of PFM1 modifier components, and ash-cement ratio.

bonds between cellulose ether macromolecules and SP-3 molecules (Fig. 3.3), since naphthalene and cellulose molecules are characterized by the presence of hydrogen atoms bonded to carbon atoms that form the rings of their molecules. In particular, it is hydrogen bonds that determine the colloidal properties of cellulose ether and the plasticizing properties of superplasticizer.

The effect of the ash-cement ratio on the water consumption of cement-ash mixtures, as well as on their viscosity, is extreme, which is maintained both when introducing single additives and double combinations. At the optimal ash-cement ratio, a lower content of superplasticizer is required to achieve the minimum possible water consumption of mixtures.

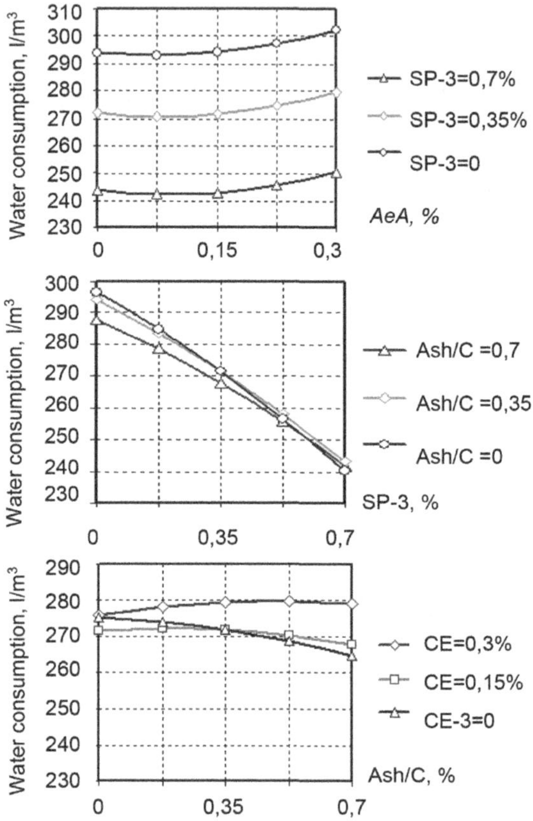

Fig. 2.2 Dependence of water demand of cement-ash mixtures on the content of PFM2 modifier components, and ash-cement ratio.

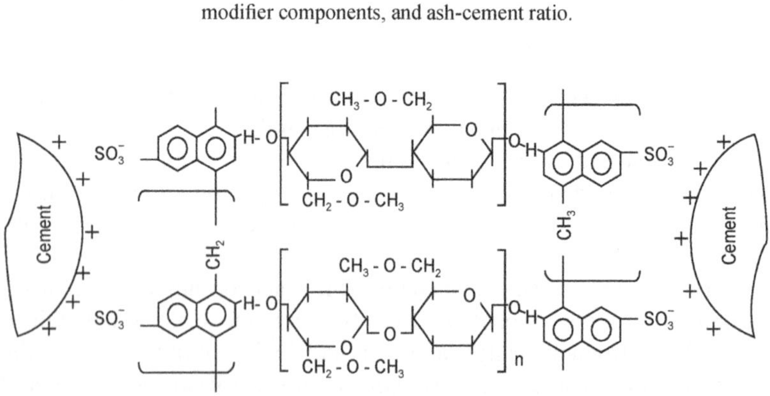

Fig. 2.3 Interaction of cement particles, naphthalene-formaldehyde superplasticizer (SP NF), and cellulose ether.

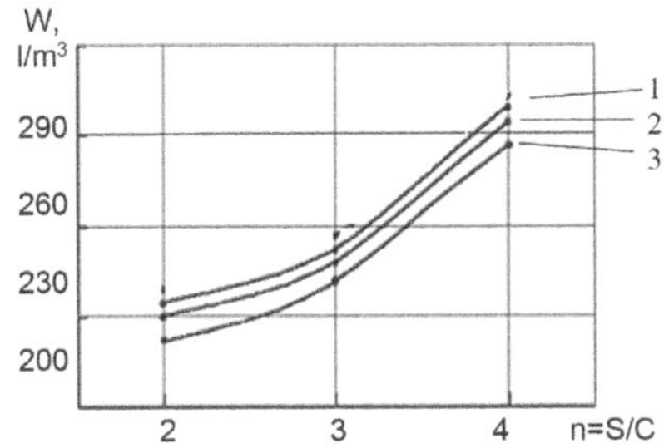

Fig. 2.4 The effect of the sand-cement ratio (n) of a mortar mixture on its water consumption (Ash/C = 0, 4): 1 – PFM$_2$ (SP-3 – 0, 3% CE-10%); 2 – PFM1 (SP-3 – 0, 3% AeA – 0, 03%); 3 – PFM 1 (SP-3 – 0, 7% AeA – 0, 03%).

The water consumption of the mortar mixes, along with the cement-ash binder, should be significantly affected by the content and water consumption of sand [30, 31]. This well-known position of concrete and mortar technology is also confirmed for modified cement-ash mixes.

Figure 2.4 shows the experimental data obtained for the water consumption of mortar mixtures of different compositions without and with PFM additives using the starting materials. The water consumption was brought to the spreading of the mortar mixture cone of 170...180 mm.

From the obtained data, it can also be concluded that the plasticizing effect of the superplasticizer SP-3 is more pronounced in 'fat' mixtures with a larger amount of cement-ash binder, and the air-entraining additive – on the contrary – in 'leaner' mixtures. If the effect of the superplasticizer is mainly due to the adsorption effect on cement and ash grains. then the air-entraining additive provides a higher content of entrained air bubbles and a corresponding decrease in the viscosity of the mixture with a higher sand content.

Water separation of mortar mixtures is the result of sedimentation compaction and precipitation of solid particles. It is determined by the water-holding capacity of the binder and its individual components, as the model (2.4) analysis shows and also depends on the water-holding capacity of aggregates and their content in the mixture. Based on modified cement-ash pastes, mortar mixtures of composition 1:2, 1:3, and 1:4 were produced at Ash/C = 0, 4 and cone spread of 170...180 mm. The water separation of mortar mixtures was determined after their settling in a cylinder.

Evaluation of the obtained data shows that, as expected, with the introduction of sand and an increase in its content in the mixture and dispersion (and, accordingly, water consumption), the water-holding capacity of the mortars increases (Figs. 2.5...2.6).

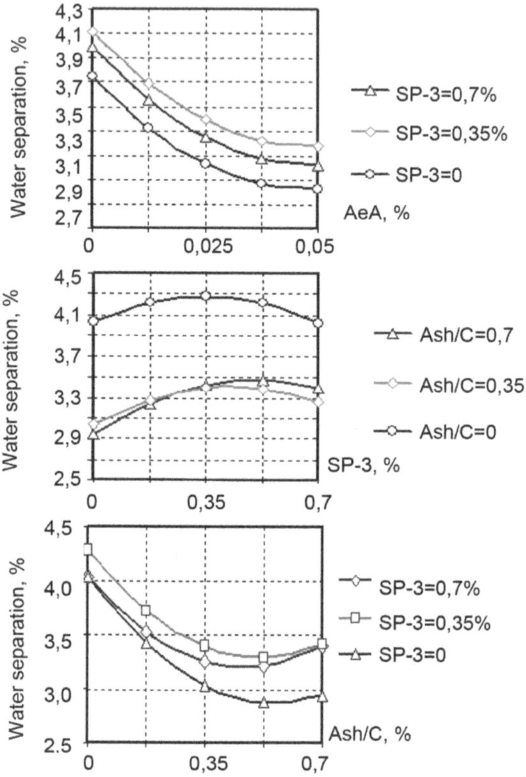

Fig. 2.5 Dependence of water separation of cement-ash mixtures on the content of the PFM_1 modifier components, and the ash-cement ratio.

The water-holding capacity of mortar mixtures is obviously affected by the volume of entrained air and especially the ability of cellulose ethers to form aquacomplexes that firmly retain water. The energy of water retention by cellulose molecules is in the range of 1250...2500 kJ/kg. As the analysis shows [30], the calculated dependence for an approximate estimate of the amount of water retained by modified cement-ash mixtures can be represented in the form of an equation (W_h):

$$W_h^c = (1{,}35 \ldots 1{,}65)K_{n.c}\,(C + \mathrm{Ash}) + W_S S + KP_{\mathrm{mod}}, \qquad (2.7)$$

where, P_{mod} – parameter, which for the studied PFM is estimated by the content AeA and CE;

$K_{n.c}$ – normal cement consistency;

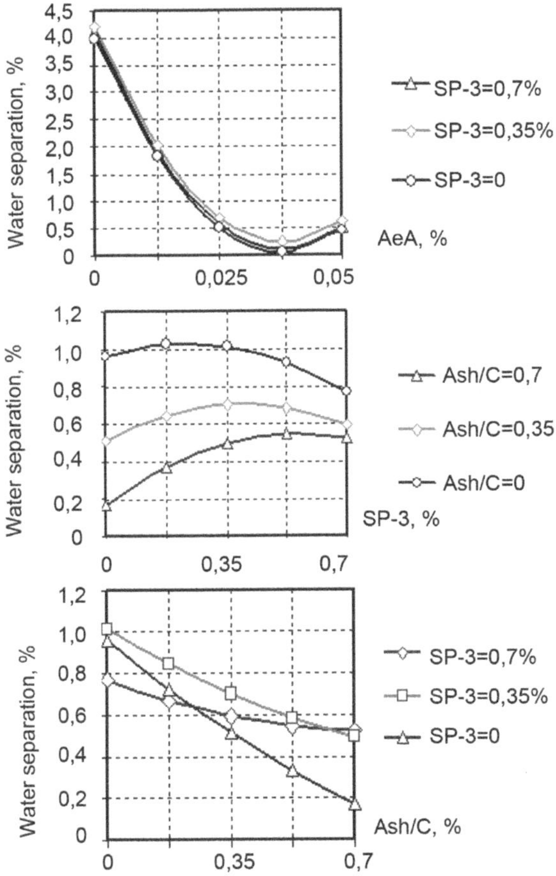

Fig. 2.6 Dependence of water separation of cement-ash mixtures on the content of the PFM_2 modifier components, and the ash-cement ratio.

W_s – water demand of sand, deter mined by the method of Bazhenev, [32];

S – sand content.

Workability. To study the influence of the entire set of composition factors of mortar mixtures filled with ash and PFM additives on their workability determined by the mortar standard cone penetration, performed algorithmic experiments according to the typical three-level plan Na_5 [22]. The experimental planning conditions are given in Table 2.3, and obtained by statistical processing of the models of mortar mixtures workability are in Table 2.4.

Table 2.3 Conditions for planning an experiment when studying the workability of mortar mixtures with PFM additives.

Technological Factors		*Levels of Variation*			*Variation Interval*
Natural View	*Coded View*	*– 1*	*0*	*+1*	
Air-entraining admixture content, % of cement mass	$X_{1(I)}$	0	0,025	0,05	0,025
Content of water-retaining additive (CE), % of cement mass	$X_{1(II)}$	0	0,15	0,3	0,15
Content of superplasticizer SP-3, % of cement mass	X_2	0	0,35	0,7	0,35
Ash/C	X_3	0	0,35	0,7	0,35
Water consumption W, kg/m^3	X_4	240	270	300	30
W/C	X_5	0,6	0,8	1,0	0,2

Table 2.4 Experimental and statistical models of the (cone penetration) of modified cement-ash mortar mixtures.

Modifier	*Regression Equations*
PFM$_1$ (superplasticizer SP-3 + air-entraining additive)	$P_{c(1)} = 9{,}73 + 1{,}08X_{1(I)} + 2{,}2X_2 + 0{,}62X_3 + 1{,}25X_4 - 1{,}14X_{1(I)}^2 + 0{,}37X_2^2 + 0{,}62X_3^2 + 0{,}37X_4^2 + 0{,}37X_5^2 + 0{,}28X_{1(I)}X_2 + 0{,}28X_2X_4 + 0{,}41X_3X_5 - 0{,}59X_4X_5$ (2.8)
PFM$_2$ (superplasticizer SP-3 + water-retaining additive)	$P_{c(2)} = 8{,}44 - 0{,}56X_{1(11)} + 2{,}47X_2 + 0{,}97X_4 + 0{,}07X_{1(11)}^2 + 0{,}82X_2^2 - 0{,}18X_3^2 - 0{,}18X_4^2 + 0{,}07X_5^2 - 0{,}25X_{1(11)}X_2 + 0{,}38X_2X_5 - 0{,}50X_3X_4 - 0{,}69X_4X_5$ (2.9)

Analysis of the obtained models shows that (Fig. 2.7) the change in the workability of modified mortars is most influenced by the content of the superplasticizer (X_2) and the water consumption (X_4). The rather significant effect of the interaction of factors $X_{1(I)}$ and X_2 is noteworthy, which shows that the simultaneous change in these factors enhances their integral effect. There is a certain interaction between factors X_3 and X_5 (ash-cement and water-cement ratio).

In all mixtures, a constant ratio of cement and sand by weight (1:3) was ensured.

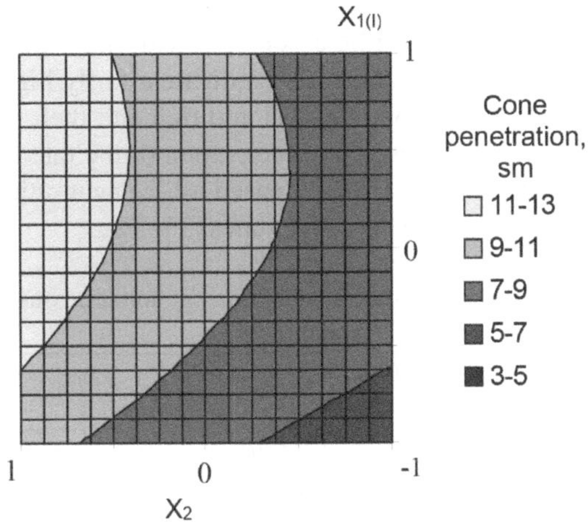

Fig. 2.7 Isolines of the cone penetration of mortar mixtures modified by PFM_1 (Ash/C = 0.35, W = 270 kg/m^3, W/C = 0.8).

In the obtained polynomial models for penetration of the cone in the mortar mixture the influence of the factor X_5 – water-cement ratio in the range of 0.6...1.0 – turned out to be insignificant. From concrete technology it is known [32] that up to a certain critical C/W, the rule of constancy of water demand is preserved, that is, with a change in C/W, water demand remains practically constant. Obviously, in this case, the rule of constant water consumption for mortar mixtures also holds for different dosages of PFM components and ash-cement ratios. In this case, along with the volume of the filled dough, the ash-cement ratio significantly affects workability.

Separate experiments were conducted to study the change in workability over time of ash-cement mixtures with PFM additives. The experiments were conducted at a temperature of 20 ± 2°C. The mixtures had a W/C ratio of 0.6, and the type and consumption of individual PFM components were varied. The initial workability by the cone penetration was 11...13 cm. The results of the experiments are given in Table 2.5. They show that the cast consistency of the mixture (the depth of the cone penetration is more than 10 cm) in mixtures with the addition of SP-3 (0.3%) without ash is preserved for 20 minutes, with the addition of SP-3 (0.3%) + AeA (0.03%) – 60 minutes, with the addition of SP-3 (0.3%) + CE (0.3%) – 100 minutes. The introduction of fly ash at Ash/C < 0.4 has a positive effect on maintaining workability, with a further increase in the ash-cement ratio, the rate of workability decline increases slightly. The positive effect of ash on stabilizing the workability of mortar mixtures in the presence of SP-3 can be associated with their increased water-holding capacity. Other authors also indicate a fairly high viability of mortar mixtures with an optimal ash addition and their suitability for long-distance transportation. The acceleration of the rate of decrease in workability of mixtures with an increased ash-cement ratio is explained by the acceleration of cement setting times [33–35].

Table 2.5 Change in workability over time of mortar mixtures with PFM additives.

Composition No.	*Ash/C*	*Workability of Mixtures by Cone Penetration, cm, after, Minutes*				
		20	*40*	*60*	*80*	*100*
Additive SP-3 (0, 3%)						
1	0	10,8	7,3	5,2	2,4	-
2	0,35	11,4	8,5	6,1	3,7	-
3	0,7	9,5	6,5	3,7	1,5	-
Additive PFM_1: SP-3 (0, 3%)+AeA 0, 03%						
4	0	12,2	10,5	9,9	6,8	3,2
5	0,35	11,6	10,6	10,2	7,5	4,5
6	0,7	11,2	10,0	8,6	5,8	3,4
Additive PFM_2: SP-3 (0, 3%)+CE 0, 3%						
7	0	11,6	10,7	10,4	10,1	6,9
8	0,35	11,5	10,6	10,5	10,3	8,1
9	0,7	10,5	10,4	10,0	8,7	6,4

Air-entraining and, especially, water-retaining additives have a stabilizing effect on the loss of workability. They allow to significantly extend the durability of mortar mixtures containing a superplasticizer.

The kinetics of strength loss by modified mortars agrees with the change in setting times and plastic strength of cement-ash dough.

The workability decline curves of mortar mixtures with PFM additives can be divided into stages of relative stability and progressive mobility decline. Both the entire period of workability decline and the period of relative stability (Fig. 2.8) are closely correlated with the period of structure formation on the plastic strength curves, other things being equal. The rate of workability decline is more affected by the addition of ash.

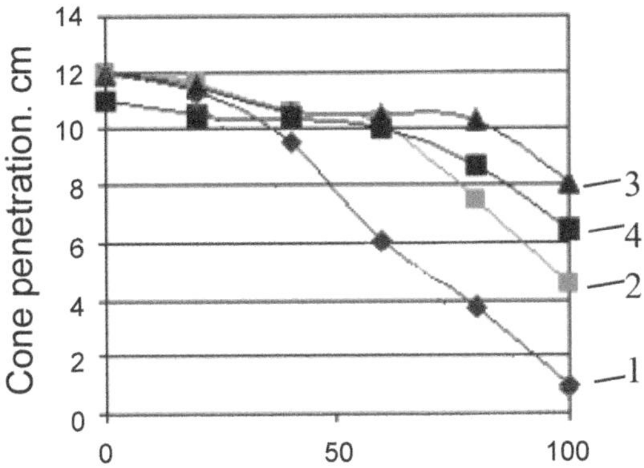

Fig. 2.8 Change in workability of mortar mixtures over time (Ash/C = 0,35): 1 – SP-3 = 0,3%; 2 – SP-3 (0,3%) + AeA 0,03%; 3 – SP-3 (0,3%) + CE 0,3%; 4 – SP-3 (0,3%) + CE 0,3% (Ash/C = 0,7).

The kinetics of changes in the rheological properties of mortar mixtures over time significantly depends on the temperature factor. At temperatures above 30°C, the period of relative stability of the workability of mixtures with PFM_1 practically disappears, although the initial workability with PFM additives is higher. At 40°C, a noticeable decrease in the initial workability of mortar mixtures is already noted: up to 6–7 cm for mixtures without additives, and up to 7–8 cm for mixtures with PFM. At 60°C, the initial workability decreases to 3–4 cm.

Analysis of experimental data shows that the workability of mortar mixtures over time generally varies according to the equation:

$$P_{c\tau} = P_{c0}\,(1 - k\tau), \qquad (2.10)$$

where, P_{co} – workability (cone penetration) of the mortar mixture, determined immediately after mixing ($\tau = 0$);

$P_{c\tau}$ – workability of the mortar mixture after time τ, h, after mixing;

k – coefficient of relative change in the penetration of the mixture, which depends on the ambient air temperature and the type of PFM.

The change in workability over time should obviously be affected by all factors that determine the rate of cement hydration. In addition to W/C and temperature, such factors include the activity and dispersion of cement, additives, and their ratio [36].

2.2 Adhesion Properties of Modified Cement-ash Mortars

Modern ideas about the mechanism of adhesion and bonding. One of the main tasks of modifying mortars for 3D printing is to increase their adhesive capacity, which contributes to improving the quality and durability of masonry.

At present, there is no generally accepted theory that satisfactorily explains the bonding process. Due to the specificity and diversity of phenomena that occur at different stages of the bonding process, the creation of a general theory of bonding is significantly complicated [37].

One of the early theories proposed to explain the bonding process is the Mc.Bain hypothesis [38], which considers this process as a mechanical 'jamming' of the adhesive substance into the pores (or recesses) of the material being bonded. However, the positions put forward by Mc.Bain were refuted in subsequent works [39]. The adsorption, electrical, diffusion, and chemical theories of adhesion have acquired the greatest importance at present.

The adsorption theory of adhesion considers the formation of a bond between an adhesive and a bonded material as a result of intermolecular forces. The importance of adsorption phenomena for the adhesion process was first pointed out [40] in 1926. However, the foundations of the adsorption theory of adhesion were developed much later by Debrouin [41] and Mc. Laren [39].

According to Mc. Laren's ideas, the formation of an adhesive layer occurs in two stages. In the first stage, the adhesive molecules move (migrate) to the substrate surface and the polar groups of the adhesive molecules approach the polar areas of the substrate, and in the second stage, sorption occurs. When the distance between the molecules of the adhesive and the substrate is sufficiently close, molecular forces (dispersive, inductive, electrostatic) begin to act, which lead to the formation of various bonds (dipole-dipole, dipole-induced dipole, hydrogen bond, etc.).

The magnitude of the adhesion of a liquid to a solid is expressed by the following equation [42]:

$$W_a = \sigma\,(1 + \cos\varphi), \tag{2.11}$$

where, W_a – adhesion work;

σ – surface tension of a liquid;

φ – the edge angle formed by the tangent to the surface of a drop and the solid surface on which the drop is placed.

The magnitude of the edge angle is inversely proportional to the adhesion. The adhesion of two solids is determined by the work that is spent on separating the solid surfaces.

The electrical theory of adhesion is based on the idea of a double electric layer, which is formed when two surfaces are in close contact.

V.V. Karas'ov, N.A. Krotova, and B.V. Deryagin [43; 44] discovered the phenomenon of electron emission during the separation of adhesive films, which confirms the electrical theory of adhesion. A relationship between electron velocity and adhesion magnitude has also been established. Adsorption phenomena are considered important because they result in a redistribution of electrons at the distribution boundary, leading to the formation of an electric double layer.

At the next stage, the electrical theory of adhesion took into account the importance of molecular interaction forces in adhesion phenomena.

The electrical theory of adhesion is not universal and has certain shortcomings [45]. From the point of view of this theory alone, it is difficult, for example, to explain the increase in adhesion as the nature of the materials being joined becomes closer to each other. If everything were reduced to the formation of a double electron layer, an inverse relationship would have to be observed, since the closer the nature of the phases in contact, the smaller the contact potential difference should be.

According to chemical theory [46], adhesion of different materials is due to the emergence of chemical bonds. Bonding can be explained by the formation of chemical bonds, for example electrovalent, or, as they are also called, ionic, hundreds of times stronger than intermolecular forces. Under the most favorable bonding conditions, a different type of chemical bond can arise – covalent.

The mechanism of adhesion can be characterized by hydrogen bonding, which is considered as a secondary valence of the hydrogen atom, which manifests itself towards negative atoms (fluorine, oxygen, nitrogen). This manifestation of additional valence by hydrogen is possible only under the condition that the main valence binds it to an atom that is also strongly negative. Despite the fact that the hydrogen bond is much weaker than a chemical bond, it can be several times stronger than intermolecular bonds.

Bond energy is highest in chemical bonds, and to a lesser extent in hydrogen, electrostatic, dispersion, and induction bonds.

None of the above theories of bonding is universal. Although none of these theories is currently favored, it can be stated that each of them contributes to the general theory of the bonding mechanism.

When forming an adhesive film, shrinkage phenomena occur, which usually lead to the emergence of internal stresses that negatively affect the strength of adhesive joints. The magnitude of shrinkage and internal stresses depends both on the chemical nature and properties of the compounds that form the basis of the adhesive composition, and on the nature and conditions of the processes that occur during the formation of the adhesive joint.

Many studies have shown that the strength of the adhesive bond increases with decreasing adhesive film thickness. A thin adhesive layer is stronger because the probability of defects in it is lower than in a thick adhesive layer.

According to the ideas [47], the failure in the adhesive layer is caused by the limit development of plastic deformations by two rigid bonded elements; this limit in thin layers is much greater than in thick ones. The influence of the thickness of the adhesive layer on the strength is associated with the simultaneous action of two factors – plastic yielding and internal stresses [48].

Along with the introduction of polymer additives, there are a number of known methods for improving the adhesive ability of cement stone while limiting its content in concrete. One of them is based on the concept of considering cement stone as microconcrete [49; 50]. In accordance with this concept, it is advisable to increase the dispersion of cement glue, ensuring its complete hydration. Cement grains with a size of more than 40 microns, which practically do not hydrate, should be rationally replaced with fillers. This concept is based on the technologies and wet grinding of cement in combination with sand and other fillers, obtaining colloidal cement glue. Cement grinding, however, has not gained widespread popularity due to high energy consumption, imperfect design of grinding units, and rapid loss of activity. The use of colloidal cement adhesive, which is obtained using vibration mills and mixers, is limited to a very narrow range of adhesive mixtures.

Adhesive strength of cement-ash mortars with PFM additives. The adhesiveness of cement-ash mortars was determined as the strength of separation from the concrete base of a sample measuring 50×50 mm, cut from ceramic tiles. The influence of water-cement (W/C) and ash-cement (Ash/C) ratios, as well as the type and content of the investigated additives, was studied using experiments algorithmized according to plan B_4. The experimental planning conditions are given in Table 2.6.

By mathematical processing of results, experimental and statistical models of adhesion strength (2.12...2.15) were obtained and are given in Table 2.7. To analyze the obtained models, two-factor graphical dependences and response surfaces of adhesion strength were constructed (Figs. 2.9...2.11).

Table 2.6 Conditions for planning an experiment when studying the adhesion properties of modified cement-ash mortars.

Technological Factors		*Levels of Variation*			*Variation Interval*
Natural View	*Coded View*	*– 1*	*0*	*+1*	
W/C	X_1	0,6	0,8	1,0	0,2
Ash/C	X_2	0	0,35	0,7	0,35
Content of superplasticizer SP-3, % of cement mass	X_3	0	0,35	0,7	0,35
Air-entraining admixture (AcA) content, % of cement mass	$X_{4(I)}$	0	0,025	0,05	0,025
Content of water-retaining additive Tylose (CE), % of cement mass	$X_{4(II)}$	0	0,15	0,3	0,15

Table 2.7 Mathematical models of adhesive strength of cement-ash mortars.

Additive	*Mathematical Models of Adhesion Strength*
PFM_1 (SP-3 + AeA)	$R_{ad,7} = 0{,}356 - 0{,}028X_1 + 0{,}035X_3 + 0{,}017X_4 - 0{,}0459X_1^2 - 0{,}0459X_2^2 - 0{,}0209X_3^2 - 0{,}0459X_4^2 - 0{,}004X_1X_2$ (2.12) $R_{ad,28} = 0{,}639 - 0{,}028X_1 + 0{,}018X_2 + 0{,}057X_3 - 0{,}026X_4 - 0{,}0792X_1^2 - 0.074X_2^2 - 0{,}0542X_3^2 - 0{,}0542X_4^2 + 0{,}004X_1X_2 + 0{,}004X_1X_4 + 0{,}011X_2X_3$ (2.13)
PFM_2 (SP-3 + CE)	$R_{ad,7} = 0{,}392 - 0{,}027X_1 + 0{,}07X_2 + 0{,}036X_3 + 0{,}034X_4 - 0{,}0574X_1^2 - 0{,}0424X_2^2 - 0{,}0224X_3^2 - 0{,}0424X_2^4 + 0{,}002X_1X_4 + 0{,}002X_3X_4$ (2.14) $R_{ad,28} = 0{,}706 - 0{,}030X_1 + 0{,}021X_2 + 0{,}060X_3 + 0{,}052X_4 - 0{,}097X_1^2 - 0.067X_2^2 - 0{,}067X_3^2 - 0{,}067X_4^2 + 0{,}006X_1X_2 + 0{,}006X_1X_4 + 0{,}016X_2X_3 + 0{,}003X_3X_4$ (2.15)

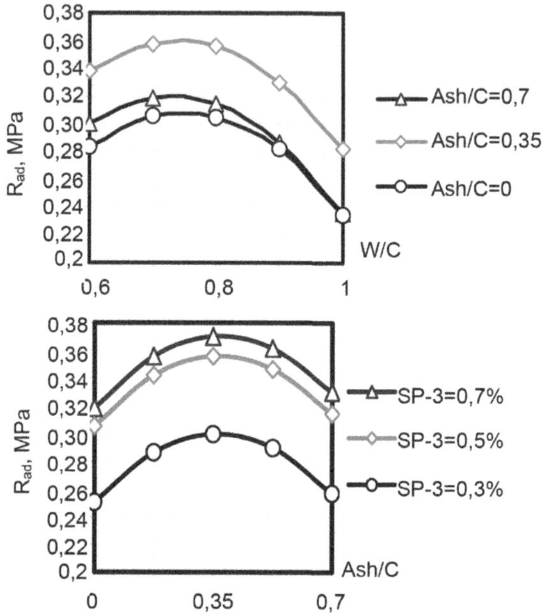

Fig. 2.9 Dependence of the adhesive strength (MPa) of cement-ash mortars modified with PFM_1 on the water-cement and ash-cement ratio at the age of 7 days.

Additionally, the influence of fly ash grinding on the adhesive strength was studied, as well as the ratio of cement-ash binder: aggregate. The results of the experiments are given in Table 2.8 and Fig. 2.12.

Analysis of the obtained adhesion strength models shows that the influence on it of both W/C and Ash/C is extreme (Fig. 2.9, 2.10). Thus, with an increase in W/C of the mortars from 0.6 to 0.7...0.75, the adhesion strength increases by 8...10%, with a further increase in W/C it decreases by 20...25% of the maximum value. At the same time, the maximum $R_{ad.}$ is observed at Ash/C = 0.35...0.4. Such an influence of the mentioned technological factors on adhesion can be explained by the influence not only of the porosity of the contact layer, but also of the degree of wetting by the base mortar.

The superplasticizer SP-3 and the air-entraining additive Mix-uk have a positive effect on the adhesion of mortars (Fig. 2.11) both as a result of changing their surface energy and as a result of changing the qualitative characteristics of the contact layer. The introduction of the hydrophilic additive SP-3 improves the characteristics of the contact layer, obviously, primarily increasing its wettability and reducing the content of excess moisture. When the content of a superplasticizer increases 0 to 0.35% of

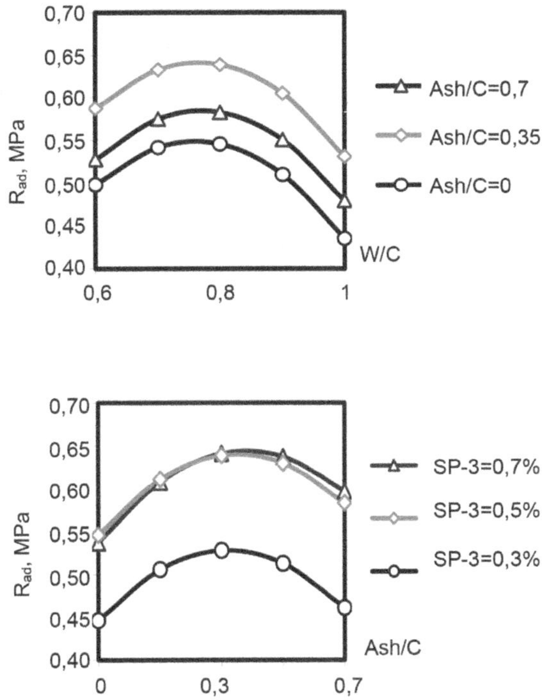

Fig. 2.10 Dependence of the adhesive strength (MPa) of cement-ash mortars modified with PFM_1 on the water and ash-cement ratio at the age of 28 days.

the cement mass, the adhesive strength increases by 16...35% at W/C = 0.6 and by 25...40% at W/C = 1.0. Further increase in the content of SP-3 to 0.7% leads to an increase in R_{ad} by 13...20%.

The introduction of the air-entraining and, especially, polymeric additive Tylose into mortars provides the necessary water-retaining capacity of the mortars' mixture and reduces the thickness of the adhesive layer, which also has a positive effect on the adhesion value. Other things being equal, increasing the content of the air-entraining additive from 0 to 0.025% of the cement mass increases the adhesive strength by 30...45%, with a further increase in the AeA content, the adhesive strength decreases slightly. Increasing the content of the water-retaining additive Tylose from 0 to 0.15% leads to an increase in adhesion strength by 25...55%, a further increase in the content of the additive has a weak effect on the adhesion of ash-containing mortars.

An additional factor that contributes to increasing the adhesive strength of mortars is the increase in the dispersion of fly ash (Table 2.8,

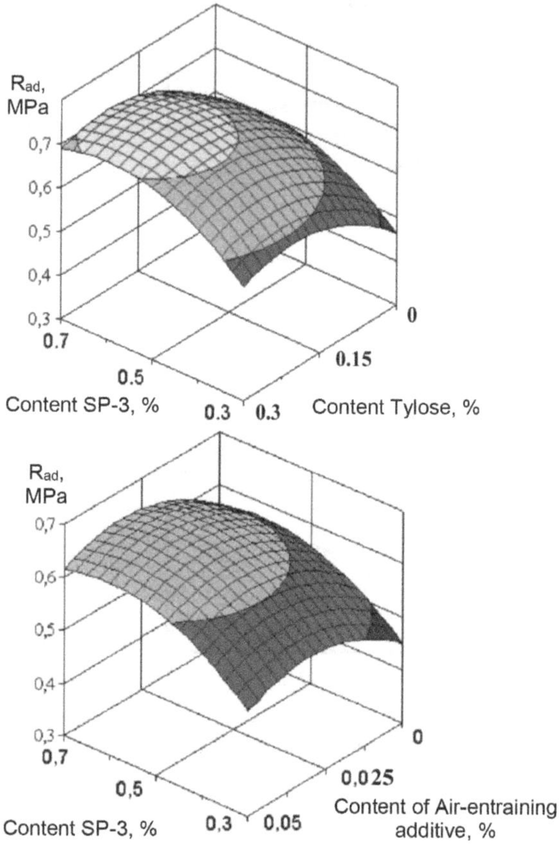

Fig. 2.11 Dependence of the adhesive strength (MPa) of cement-ash mortars at the age of 28 days on the content of additives-modifiers.

Fig. 2.12). Thus, grinding the ash to a specific surface area of 390 m^2/kg allows to increase the adhesion strength of the mortars to the base at the age of 7 days by 14...26%, at the age of 28 days – by 13...18%.

It is known that new, freshly formed surfaces of mineral materials have significantly higher surface energy values, which determines their higher adhesive activity [51]. Mechanical processes during grinding of mineral and organic materials cause, along with an increase in their surface energy, an increase in the isobaric potential of powders and, accordingly, their chemical activity, which also contributes to high adhesive strength when they come into contact with a binder. However, it should be taken into account that ground powders are prone to rapid deactivation. The duration of existence in the air of radicals that arise during mechanical and chemical processing is only $10^{-3}...10^{-6}$ sec. Adsorption of water vapor and carbon dioxide from the air by freshly ground powders and saturation of uncompensated molecular forces leads not only to 'aging' of the filler surface, but also serves as an additional obstacle to the formation of reliable adhesive contacts. Because of this, the mechanochemical activation of fillers is effective in creating a primary contact layer of structured binder on their grains directly during the grinding process [52].

Table 2.8 Adhesive strength of modified cement-ash mortars using ash of different dispersion.

(C+Ash)/Sand	*Specific Surface, m²/kg*	*Adhesive Strength, MPa, at Age*	
		7 days	*28 days*
Modifier PFM$_1$ (W/C = 0, 8; Ash/C = 0, 6; SP-3 = 0, 5%; AeA = 0, 03%)			
1 : 3	290	0,34	0,60
	340	0,38	0,66
	390	0,43	0,71
1 : 4, 5	290	0,28	0,53
	340	0,32	0,57
	390	0,35	0,62
Modifier PFM$_2$ (W/C = 0, 8; Ash/C = 0, 6; SP-3 = 0, 5%; CE = 0, 3%)			
1 : 3	290	0,38	0,64
	340	0,42	0,69
	390	0,45	0,73
1 : 4, 5	290	0,29	0,55
	340	0,31	0,59
	390	0,33	0,62

2.3 Strength and Deformation Properties of Cement-ash Mortars

Strength of cement-ash mortars with PFM additives. The operational reliability of mortars, for 3D printing along with their adhesion, is characterized by the value of cohesion, which determines the strength and durability indicators of the mortar.

The strength studies of modified mortars were performed by implementing algorithmized experiments, the planning conditions of which are given in Table 2.6. Beam samples that hardened at normal

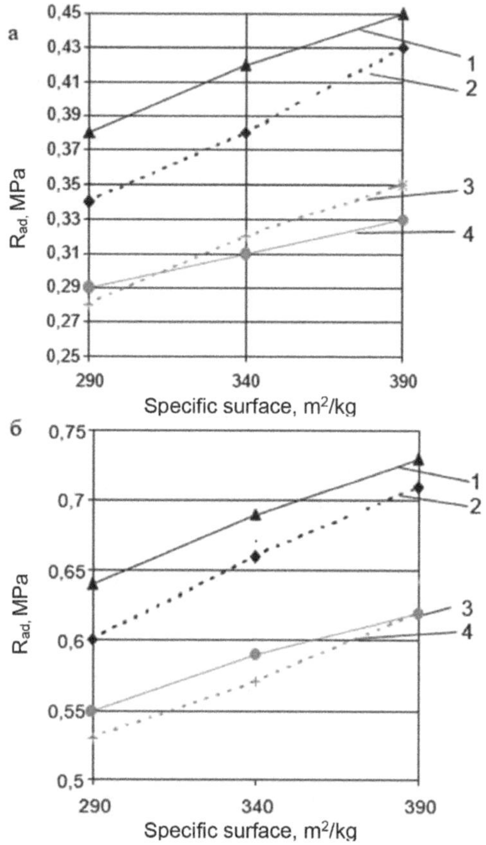

Fig. 2.12 Dependence of the adhesive strength of modified cement-ash mortars on the dispersion of ash and the ratio of binder: sand: (a) at the age of 7 days, (b) at the age of 28 days: 1, 2 – (C+Ash): S = 1:3; 3, 4 – (C+Ash): S = 1:4, 5; 1, 4 – CE = 0, 15%, sand S_1; sand S_2; 2, 3 – AeA = 0, 025%, sand S.

conditions were tested at the age of 28 days. Statistical processing of the experimental results yielded regression equations for compressive strength (R_{comp}) and bending strength (R_{bend}), given in Table 2.9, and graphical dependencies illustrating them are in Fig. 2.13… 2.16.

Analysis of the obtained equations and graphs shows that in the studied cement-ash mortars the decisive influence on both compressive and flexural strength is exerted by W/C. The next most important factor is the content of the superplasticizer. The ash-cement ratio also plays a significant role, the influence of which on both compressive and flexural

Table 2.9 Mathematical models of the strength of modified cement-ash mortars.

Additive	*Regression Equation*
PFM$_1$ (SP-3 + AeA)	$R_{comp} = 24{,}8 - 8{,}04X_1 + 1{,}207X_2 + 2{,}49X_3 - 2{,}03X_4 + 1{,}04X_1^2 - 0{,}458X_2^2 + 0{,}342X_3^2 + 2{,}71X_4^2 + 0{,}33X_1X_2 + 0{,}66X_1X_3 + 0{,}23X_2X_3 + 0{,}44X_2X_4 - 0{,}18X_3X_4$ (2.16) $R_{bend} = 3{,}74 - 0{,}986X_1 + 0{,}353X_3 - 0{,}252X_4 + 0{,}401X_1^2 - 0{,}151X_2^2 - 0{,}051X_3^2 - 0449X_4^2 - 0{,}044X_1X_4 + 0{,}056X_1X_4$ (2.17)
PFM$_2$ (SP-3 + CE)	$R_{comp} = 25{,}5 - 8{,}4X_1 + 1{,}106X_2 + 2{,}61X_3 - 1{,}62X_4 + 1{,}83X_1^2 - 1{,}62X_2^2 + 1{,}03X_3^2 - 1{,}87X_4^2 + 0{,}32X_1X_2 + 1{,}77X_1X_3 + 1{,}21X_2X_3 - 1{,}18X_3X_4$ (2.18) $R_{bend} = 3{,}93 - 0{,}935X_1 + 0{,}017X_2 - 0{,}291X_3 - 0{,}289X_4 + 0{,}459X_1^2 - 0{,}267X_2^2 + 0{,}009X_3^2 - 0491X_4^2 - 0{,}069X_1X_2 - 0{,}044X_1X_4 - 0{,}0313X_3X_4$ (2.19)

strength is extreme. For all studied compositions, the ratio of ash/cement close to the optimum is Ash/C = 0, 35...0, 4.

The positive effect of ash addition on strength in a number of works [53, 54] is characterized by "cementing efficiency" – cement savings, which are achieved when 1 kg of ash is added to the mortars at a constant W/C.

For ash-cement systems, a compressive strength formula has been proposed [53], which is a modification of the well-known Bolomey-Skramtaev formula:

$$R_{comp} = AR_C\left(\frac{C + K_{c,e}Ash}{W}\right), \quad (2.20)$$

where, $K_{c.e}$ – cementing efficiency coefficient.

As follows from the obtained data, the value of $K_{c.e}$ will significantly depend on the adopted ash-cement ratio, and at a given cement consumption – on the ash consumption. Obviously, as follows from Fig. 2.13, 2.14, 2.15, the positive effect of ash on the strength of mortars can be significantly reduced or reduced to zero with a not optimal cement-ash ratio and ash consumption.

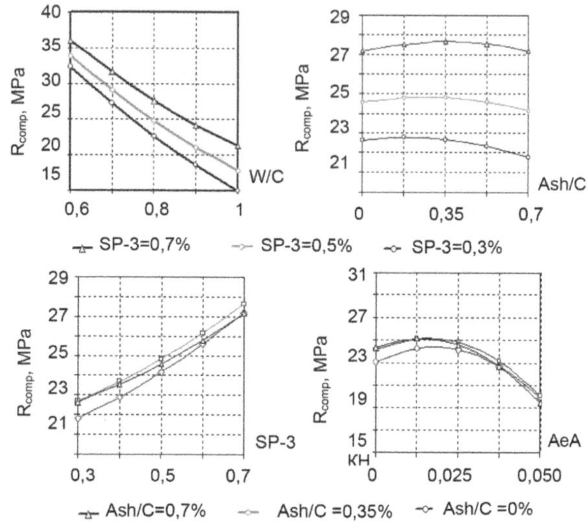

Fig. 2.13 Dependence of compressive strength of cement-ash mortars modified with the PFM-1 additive on technological factors.

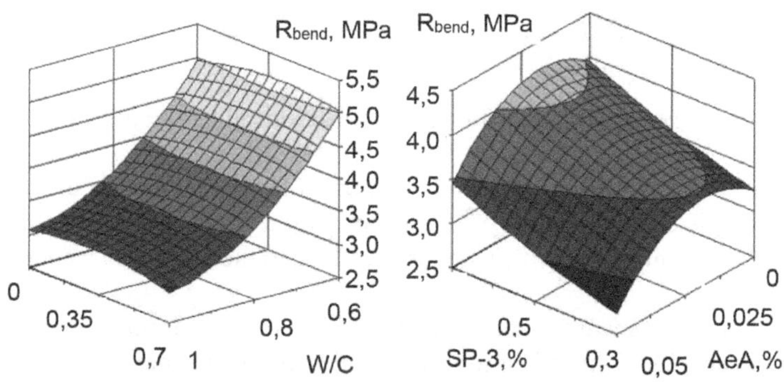

Fig. 2.14 Dependence of the bending strength of cement-ash mortars modified with the PFM1 additive on technological factors.

The influence of additives – polyfunctional modifiers (PFM) on the strength of mortars depends on their composition and content [55]. With constant workability, an increase in the content of the superplasticizer SP-3 in the composition of the PFM naturally leads to an increase in strength. In particular, with an increase in the content of SP-3 from 0.3 to 0.7%, the compressive strength of the mortars increases by 23...34%, and the bending strength – by 18...24%, all other things being equal. An

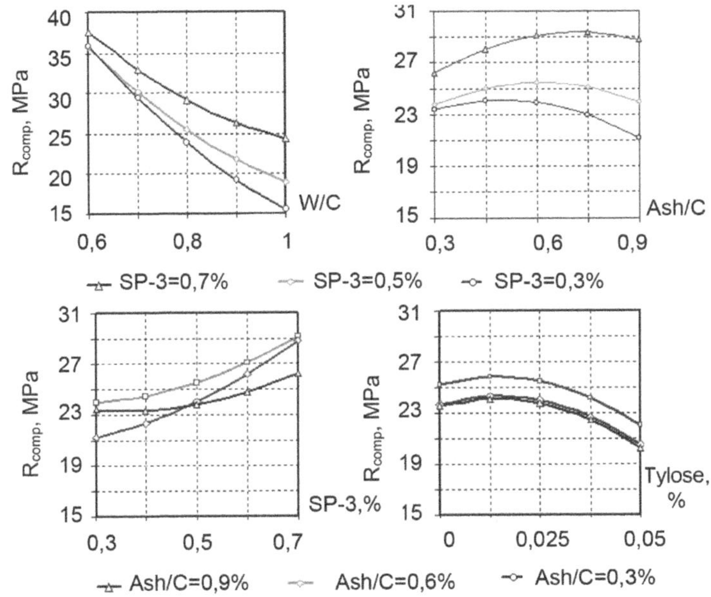

Fig. 2.15 Dependence of compressive strength of cement-ash mortars modified with PFM2 additive on technological factors.

increase in the content of water-retaining and, especially, air-entraining additives leads to a decrease in strength by 10...14% in the first case and by 14...21% in the second. The most noticeable decrease in strength is observed when the content of Tylose additives exceeds 0.3%, and Mix-uk additives exceed 0.04% of the cement mass. However, when assessing the impact of these additives, it is worth considering the fact that when laying on porous substrates, the negative impact of these additives on the strength of modified mortars is leveled out by the increased workability of mortar mixtures and the high quality of the mortar joint.

Modification of mortars with PFM additives has a positive effect on the ratio of tensile and compressive strengths, which is one of the criteria for the crack resistance [56]. Figure 2.17 shows data illustrating the change in the ratio R_{bend}/R_{comp} at an age of up to 180 days for mortars modified with PFM. As follows from Fig. 2.17, modification of mortars with organic and organo-mineral additives has a positive effect on the given crack resistance criterion. An increase in its values is characteristic of modified mortars both at the age of 28 days and at other hardening times.

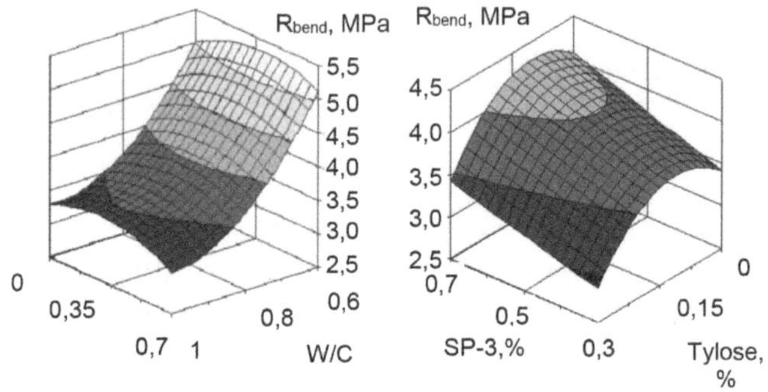

Fig. 2.16 Dependence of the bending strength of cement-ash mortars modified with the PFM2 additive on technological factors.

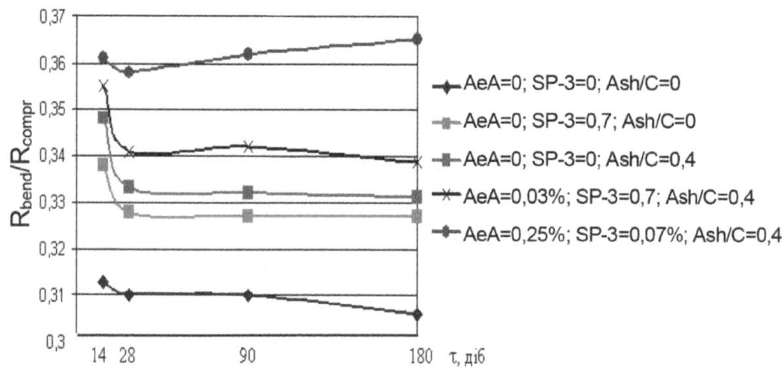

Fig. 2.17 Dependence of the crack resistance criterion (R_{bend}/R_{comp}) on the hardening time of the mortar modified with PFM_1 and PFM_2 (W/C = 0.7).

Deformational properties of mortars with modifying additives. The value of the elastic modulus was calculated using the formula:

$$E_e = \sigma/\varepsilon_e, \quad (2.21)$$

where, σ – applied stresses;

ε_e – longitudinal relative deformations.

The values of the strength and modulus of elasticity of the investigated mortars are given in Table 2.10. The cement:sand ratio is taken from the condition of achieving necessary workability by the mortar, which corresponds to the penetration of the cone by 8...10 cm.

Table 2.10 The effect of polyfunctional modifiers on the prismatic strength and elastic modulus of solutions.

No.	*Water-cement Ratio*	*Ash-cement Ratio*	*Content of AeA or CE, % of the Binder Mass*	*Content of SP-3, % of the Binder Mass*	*Compressive Strength, MPa, at the Age of 28 days*		*Modulus of Elasticity $E_e \times 10^4$ MPa*
					Cubic	*Prismatic*	
Cement-ash mortar with PFM_1 (SP-3+AeA)							
1	0,6	-	-	-	26,9	22,7	2,11
2	0,6	-	-	0,7	28,7	24,2	2,26
3	0,6	0,4	-	-	31,3	27,2	2,45
4	0,6	0,4	0,03	0,7	29,8	25,0	2,33
5	0,7	-	-	-	24,9	20,1	2,11
6	0,7	-	-	0,7	26,2	22,2	2,16
7	0,7	0,4	-	-	27,8	24,7	2,28
8	0,7	0,4	0,03	0,7	25,7	22,2	2,04
9	0,8	-	-	-	23,0	19,4	2,00
10	0,8	-	-	0,7	25,6	21,7	2,07
11	0,8	0,4	-	-	26,4	22,6	2,21
12	0,8	0,4	0,03	0,7	24,9	21,2	2,03
Cement-ash mortar with PFM_2 (SP-3+CE)							
13	0,6	0,4	0,25	0,7	32,9	28,0	7,3
14	0,7	0,4	0,25	0,7	30,6	26,2	8,0
15	0,8	0,4	0,25	0,7	28,5	24,0	8,5

Analysis of the data in Table 2.10 shows that the prismatic strength of the investigated mortars lies within 0.81...0.88 of R_{comp} and can be approximately found from the equation:

$$R_{pr} = 0{,}85\ R_{comp}. \tag{2.22}$$

As is known, for heavy-weight concrete , the correlation dependence $R_{pr} = 0.783 R_{comp}$ is valid [54]. Higher ratios of prismatic and cubic strength have been recorded for mortars and sand concrete by other authors [57].

When predicting the modulus of elasticity of concrete when loaded at age τ, the most commonly used dependences [58] are:

$$E_e = \frac{E_m R_{comp.\tau}}{S + R_{comp.\tau}}, \tag{2.23}$$

where, $R_{comp.\tau}$ – cubic compressive strength of concrete;
E_m and S – empirical constants. The building normie recommends the values $E_m = 52000$; $S = 23$.

Analysis of the experimental values of the modulus of elasticity of mortars (Table 2.10) shows that they are approximately 20% lower than those calculated by formula (2.23).

A comprehensive assessment of the deformability of concrete and its crack resistance is the ratio of the tensile or bending strength to the static or dynamic modulus of elasticity. These parameters are quite close. As shown in [55], for fine-grained concrete, the formulas for calculating the tensile strength at splitting (R_{sp}) and the tensile strength at bending (R_{bend}) differ only in the coefficients:

$$R_{sp} = 0{,}57 R_{comp}^{2/3}; \quad R_{bend} = 0{,}99 R_{comp}^{2/3}. \tag{2.24}$$

Comparison of experimental values of static and dynamic elastic moduli shows that the ratio E_{st}/E_{dyn} for concretes and mortars is in the range of 0.87...0.95. At the same time, lower values are characteristic of materials with a compressive strength of less than 25 MPa. The ratio E_{st}/E_{dyn} is close to the value of ultimate elongation and is called the conditional extensibility ($\varepsilon_{c.e}$). It is not difficult to show that

$$\varepsilon_{c.e} = \frac{R_{sp}}{E_{dyn}} \approx \frac{0{,}57 R_{bend}}{0{,}99 \cdot 1{,}1 E_{st}} \approx 0{,}52 \frac{R_{sp}}{E_{st}}. \tag{2.25}$$

Table 2.11 shows the parameter E_e and $\varepsilon_{c.e}$ values for the tested mortars at the age of 28 days, calculated by the formulas (2.23) and (2.25).

The conditional extensibility of unmodified mortars tends to decrease with increasing curing time, while it increases for modified mortars. Obviously, the sign of the change in the conditional extensibility of the mortar and, accordingly, the crack resistance with increasing age is associated with the nature of the change in the ratio of tensile strength (R_{str}) and compressive strength (R_{comp}). For mortars with PFM additives, this ratio is higher than for unmodified mortars both at 28 days and at subsequent hardening times. A higher ratio of tensile strength to compressive strength for concretes and mortars with active mineral additives is also noted in the works of other researchers [32].

Shrinkage of prism samples with dimensions of 40 × 40 × 160 mm of mortars with PFM additives was measured using a tripod device equipped with a watch-type indicator with a division value of 0.001 mm. The benchmarks in the end faces of the prisms were installed 8 hours after their compaction in the molds. The samples were stored at a temperature of 20 ± 2°C and a relative humidity of 75 ± 5%. The shrinkage deformation curves are shown in Fig. 2.18.

Table 2.11 Conditional extensibility of modified mortars.

No	*Water-cement Ratio*	*Ash-cement Ratio*	*AeA or CE, % of Binder Mass*	*SP-3, % of Binder Mass*	*Bending Strength, MPa*		*Modulus of Elasticity $E_e \times 10^4$ MPa*	*Conditional Extensibility, $\varepsilon_{c.e} \times 10^{-4}$*
Cement-ash mortar with PFM_1 (SP-3+AeA)								
1	0,6	-	-	-	8,34	0,31	2,11	2,06
2	0,6	-	-	0,7	9,76	0,34	2,26	2,25
3	0,6	0,4	-	-	10,02	0,32	2,45	2,13
4	0,6	0,4	0,03	0,7	9,83	0,33	2,33	2,19
5	0,7	-	-	-	7,72	0,31	2,11	1,90
6	0,7	-	-	0,7	8,65	0,33	2,16	2,08
7	0,7	0,4	-	-	9,17	0,33	2,28	2,09
8	0,7	0,4	0,03	0,7	8,74	0,34	2,04	2,23
9	0,8	-	-	-	7,36	0,32	2,00	1,91
10	0,8	-	-	0,7	8,70	0,34	2,07	2,19
11	0,8	0,4	-	-	8,71	0,33	2,21	2,05
12	0,8	0,4	0,03	0,7	8,96	0,36	2,03	2,30
Cement-ash mortar with PFM_2 (SP-3+CE)								
13	0,6	0,4	0,25	0,7	11,84	0,36	2,57	2,40
14	0,7	0,4	0,25	0,7	11,02	0,36	2,44	2,35
15	0,8	0,4	0,25	0,7	9,98	0,35	2,34	2,22

Their analysis shows that for all studied compositions shrinkage deformations stabilize up to the age of 100 days. The highest shrinkage values are found in mortars without PFM additives. Their shrinkage before damping varies in the range of 0.7...1.5 mm/m, typical for fine-grained concretes at similar W/C. In the early stages of hardening, the addition of fly ash increases shrinkage deformations, which are then equalized with the deformations of mortars without ash. The addition of PFM_2 reduces the limiting shrinkage deformations by 20...35%, the addition of PFM_1 – by 30...50%. The degree of reduction in shrinkage deformations increases as the plasticizing effect of PFM increases. The most significant reduction in shrinkage is observed with the introduction of an air-entraining additive – by 40...60%.

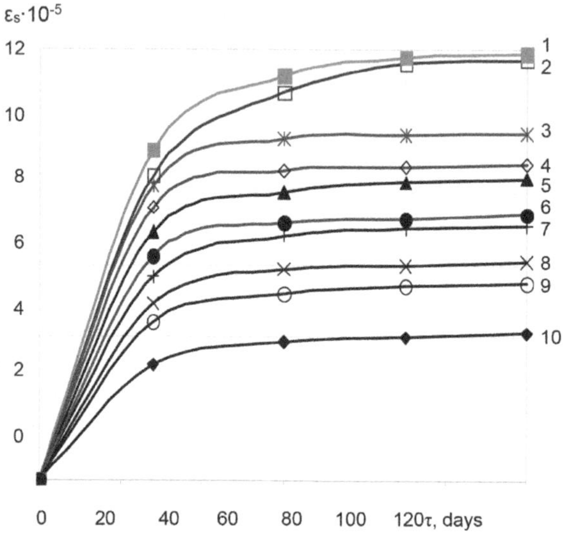

Fig. 2.18 Shrinkage deformations of modified cement-ash mortars. 1– 5 – W/C = 0, 8: 1 – mortar without PFM, Ash/C = 0, 4; 2 – mortar without PFM, Ash/C = 0; 3 – Ash/C = 0, 4, SP-3 = 0, 7%, CE = 0, 1%; 4 – Ash/C = 0, 4, SP-3 = 0, 7%, CE = 0, 4%; 5 – Ash/C = 0, 4, SP-3 = 0, 7%, AeA = 0, 01%; AeA = 0, 05%; 6– 10 – W/C = 0, 6:L 6 – mortar without PFM, Ash/C = 0, 4; 7 – mortar without PFM, Ash/C = 0; 8 – Ash/C = 0, 4, SP-3 = 0, 7%, 9 – Ash/C = 0, 4, SP-3 = 0, 7%, CE = 0, 4%; 10 – Ash/C = 0, 4, SP-3 = 0, 7%, AeA = 0, 04%.

2.4 Corrosion Resistance of Modified Mortars

Water absorption, waterproof, and sulfate resistance of mortars with PFM additives. The water absorption of mortars largely characterizes their capillary absorbency, waterproofing, and corrosion resistance.

The influence of modifiers and the ash-cement ratio on water absorption was studied on samples – prisms 40 × 40 × 160 mm at the age of 28 days. The results of the experiments are shown in Fig. 2.19. They show that for all the mortars studied, increasing the Ash/C ratio to 0.4 does not lead to a significant increase in water absorption, while with a further increase in the Ash/C ratio, water absorption increases markedly. At the same time, with an increase in the plasticizing effect of PFM, the intensity of the increase in water absorption at Ash/C > 0.6 becomes smaller, the adverse effect of high Ash/C is smoothed out.

The change in water absorption of mortars with PFM additives correlates with compressive strength, which is explained by the close dependence of both properties on the open porosity of concrete.

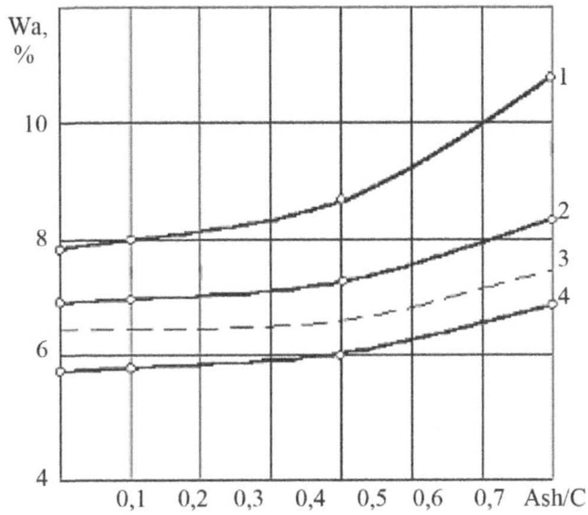

Fig. 2.19 The influence of the ash-cement ratio on the water absorption of mortars (W/C = 0.6): 1 – mortars without PFM additives 2 – PFM_2 (SP-3 = 0, 7%; CE = 0, 25%); 3 – PFM_1 (SP-3 = 0, 7%; AeA = 0, 03%); 4 – PFM_1 (SP-3 = 0, 7%;AeA = 0, 05%).

The total impact of both the qualitative characteristics of the components of mortar mixtures and their quantitative ratio on corrosion resistance is largely manifested in the water permeability of the mortars.

The waterproofing of the mortars was determined by the method of Ukrainian Standard "wet spot" on the mortars' disks in accordance with the requirements. The sulfate resistance of the mortars was determined by the change in bending strength in a 5% Na_2SO_4 solution.

The measurement results are given in Table 2.12. They are generally consistent with the previously presented data on water absorption and porosity parameters of mortars.

As is known [58], the waterproofing of the concretes and mortars is particularly reduced by sedimentation pores, which are water filtration pathways. Sedimentation processes in the mortar mixture and the water separation caused by them are quite sensitive to the content of dispersed filler, including ash.

Many studies have shown that waterproofing increases much more than strength with age of the mortar, provided that drying out during hardening is prevented. This "self-compacting" effect is associated with the filling of part of the pores with a gel during hydration, the volume of which is several times greater than the volume of unhydrated cement. The coefficient of increase in waterproofing of the concretes aged 90 and 360 days on Portland cement is 2...3 and 3...4, respectively, on pozzolanic Portland cement 3...4 and 5...6. For modified mortars, according to our data, these coefficients turned out to be lower, although quite significant (Table 2.12).

The assessment of the corrosion resistance of concrete and mortars by the change in compressive strength, while acceptable in the study of corrosion of the first and second types, is undesirable as a criterion for corrosion of the third type [49]. This is due to the fact that in the first stage of this corrosion, the compressive strength may even increase slightly due to the compaction by corrosion products.

According to the well-known classification, an environment is considered highly aggressive if, during the operation of structures in it, the decrease in strength reaches more than 20% during the year, moderately aggressive – 5...20%, weakly aggressive – less than 5%, and non-aggressive – 0%. From the data given in Table 2.12, it is clear that 5% aqueous Na_2SO_4 solution turned out to be a highly aggressive medium for the mortars without filler. For mortars filled only with ash, it turned out to be weakly aggressive, and for modified solutions it was non-aggressive. For them, there is no decrease in strength in the 5% Na_2SO_4 solution, but there is some increase.

The positive effect of the SP additive on the waterproofing of the studied concretes is naturally explained by its strong plasticizing effect.

Frost resistance. To study the frost resistance of mortars, algorithmized experiments were performed according to plan B_4. The planning conditions are given in Table 2.13.

The kinetics of the change in the strength of the mortar during cyclic freezing and thawing was determined by testing the control and main samples with the determination of the frost resistance coefficient. The tests were carried out every 25 freezing and thawing cycles [59]. As a result of statistical processing of the obtained experimental data, polynomial models of frost resistance of modified mortars were obtained (Table 2.14). Their graphical analysis is presented in Fig. 2.20...2.21.

Analysis of the obtained data shows that the frost resistance grade of the studied mortars ($K_{fr.r} \geq 0, 95$) varies in the range of F25...F250. The highest frost resistance is achieved by mortars containing an air-entraining additive. The rate of strength loss of such mortars during freezing and thawing was also lower than for mortars without the additive. The additions of SP-3 and cellulose ether have a positive effect on the frost resistance of mortars, although to a lesser extent. For filled mortars with a minimum content of organic additives and a low cement consumption (W/C=1.0), frost resistance was found to be within 25...75. Mortars with the highest ash:cement ratio have lower frost resistance.

Table 2.12 Waterproof and sulfate resistance of modified mortars.

No. Composition	*Water-cement ratio*	*Ash-cement ratio*	*Content of AeA or CE, of the Binder Mass*	*Content of SP-3, % of the Binder Mass*	*Waterproof, MPa, after, Days*			*Bending Strength in 5% Na_2SO_4 Solution at the Age of 180 days, % of the Initial*
					28	*180*	*360*	
Cement-ash mortar with PFM_1 (SP-3+AeA)								
1	0,7	-	-	-	0,3 (100%)	0,7 (233%)	0,9 (300%)	0,78
2	0,7	-	-	0,7	0,5 (100%)	0,9 (180%)	1,2 (240%)	0,92
3	0,7	0,4	-	-	0,4 (100%)	0,9 (225%)	1,3 (325%)	0,96
4	0,7	0,4	0,03	0,7	0,6 (100%)	1,2 (200%)	1,5 (250%)	1,02
Cement-ash mortar with PFM_2 (SP-3+CE)								
5	0,7	0,4	0,1	0,7	0,6 (100%)	1,3 (217%)	1,8 (300%)	1,04
6	0,7	0,4	0,4	0,7	0,7 (100%)	1,4 (200%)	1,9 (271%)	1,05

Table 2.13 Conditions for planning an experiment when studying the frost resistance of cement-ash mortars modified with organic additives.

Technological Factors		*Levels of Variation*			*Variation Interval*
Natural view	*Coded view*	*– 1*	*0*	*+1*	
W/C	X_1	0,6	0,8	1,0	0,2
Ash/C	X_2	0	0,35	0,7	0,35
Content of superplasticizer SP-3, % of the binder mass	X_3	0	0,35	0,7	0,35
Air entraining additive content (AeA), % of the binder mass	$X_{4(1)}$	0	0,025	0,05	0,025
Water-retaining additive content (CE), %	$X_{4(2)}$	0	0,15	0,3	0,15

Table 2.14 Mathematical models of frost resistance of modified mortars.

Type of PFM	*Mathematical Models of Frost Resistance*
SP-3+ AeA	$F = 149{,}9 - 67{,}1X_1 - 4{,}46X_2 + 11{,}6X_3 + 28{,}7X_{4(1)} - 36{,}6X_1^2 - 2{,}95X_2^2 + 1{,}95X_3^2 + 3{,}95X_{4(1)}^2 + 1{,}125X_1X_2 - 11{,}6X_1X_{4(1)} - 2{,}37X_2X_{4(1)} + 2{,}75X_3X_{4(1)}$ (2.26)
SP-3+ CE	$F = 135{,}8 - 69{,}4X_1 - 2{,}39X_2 + 10{,}7X_3 + 7{,}93X_{4(2)} - 33{,}4X_1^2 + 2{,}06X_2^2 + 2{,}06X_3^2 + 2{,}06X_{4(2)}^2 - 0{,}688X_1X_2 - 0{,}688X_1X_3 - 5{,}44X_1X_{4(2)} + 0{,}688X_3X_{4(2)}$ (2.27)

To explain the influence of the investigated additives on the frost resistance, the volume of entrained air was determined by the compression method for typical compositions. The measurement results are given in Table 2.15.

The obtained data show that in filled mortar mixtures containing 0.7% SP-3, air entrainment reaches 2%.

The smallest amount of entrained air was found in mixtures with the maximum content of fly ash. The introduction of an air-entraining additive increases the content of entrained air in mortars to 5...7% and significantly increases frost resistance.

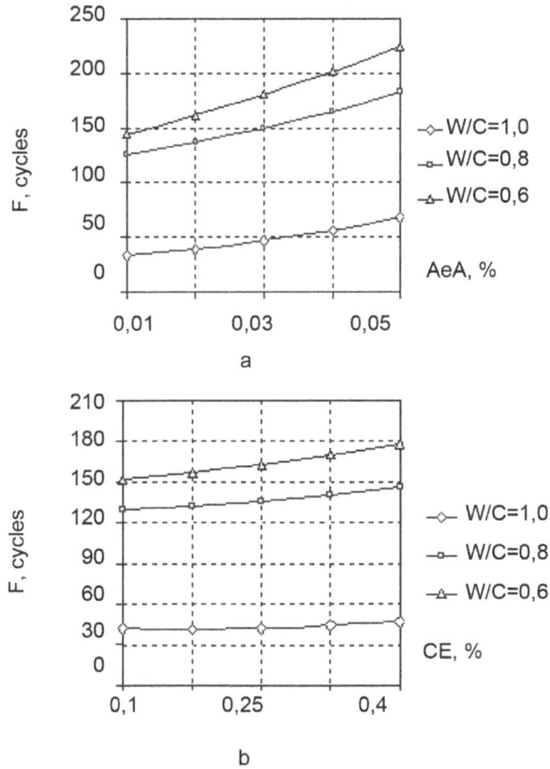

Fig. 2.20 Dependence of frost resistance of modified mortars on the content of: a – air-entraining additive; b – cellulose ether.

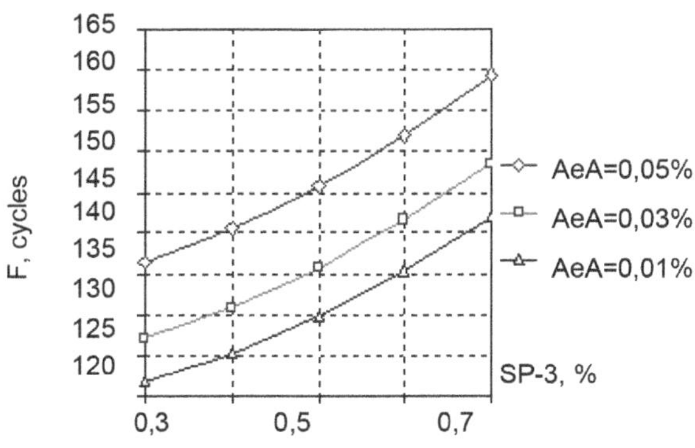

Fig. 2.21 Dependence of frost resistance of modified mortars on the content of superplasticizer SP-3.

Table 2.15 Entrained air content in modified mortars of typical compositions.

Composition of the Mortar Mixture					*Volume of Entrained Air in, %*
W/C	*Ash/C*	*Additive Content*			
		SP-3, %	*AeA, %*	*CE, %*	
1,0	0,4	0,3	0,01	-	0,7
1,0	0,7	0,3	0,01	-	0,6
1,0	0,4	0,7	0,05	-	3,8
1,0	0,4	0,7	-	0,4	2,1
1,0	0,1	0,5	-	-	1,5
1,0	0,7	0,5	-	-	0,9
0,6	0,4	0,3	0,01	-	1,4
0,6	0,4	0,7	0,05	-	5,3
0,6	0,4	0,7	-	0,4	2,0
0,6	0,1	0,5	-	-	1,9
0,6	0,7	0,5	-	-	1,1

With the same water content of mortar mixtures, the reduction of the volume of capillary pores may be associated with an increase in cement consumption and the degree of its hydration. In the considered compositions, it is probably necessary to take into account the total consumption of cement, ash and PFM, that is, the total consumption of the binder that interacts with water. As shown earlier, PFM not only increases the content of the binder in the mortar, but also provides its higher degree of hydration. Mortars with PFM additives, are also characterized by a higher ratio of the total pore volume to the volume of open pores, which should also contribute to increased frost resistance.

2.5 Optimization of Technological Parameters for Obtaining Modified Cement-ash Solutions

Features of designing mortar compositions. The most important quality indicators mortars for 3D printing are compressive strength, tensile strength, adhesive strength, frost resistance, workability, and the main compositional factors that determine them are the water-cement and ash-cement ratio, the sand: cement ratio, type, and the content of modifier additives. The required quality indicators of mortar mixes can be obtained with different ratios of these factors. Therefore, when designing the composition of the mixture, it is necessary to choose such a ratio of components that ensures not only the required properties of the mortar, but also the minimum cost of the mix.

In the most general form, the design of building mortars can be formulated as a system of equations:

$$y_1 = f_1(P_1, P_2, \dots P_{n-1}) \geq y_1^0; \quad (\text{or} < y_1^0)$$
$$y_2 = f_2(P_1, P_2, \dots P_{n-1}) \geq y_2^0; \quad (\text{or} < y_2^0)$$
$$\dots\dots \; \dots\dots \; \dots\dots \; \dots\dots \; \dots\dots \; \dots\dots \; \dots\dots \; \dots\dots \; \dots\dots;$$
$$y_{n-1} = f_{n-1}(P_1, P_2, \dots P_{n-1}) \geq y_{n-1}^0; \quad (\text{or} < y_{n-1}^0)$$
$$Cr = f(P_1, P_2, \dots P_{n-1}) \rightarrow opt, \tag{2.28}$$

where, y_1^0, y_2^0, y_{n-1}^0 – given indicators of the properties of the mixtures and mortars based on its;

$P_1, P_2 \dots P_{n-1}$ – mixture composition factors;

Cr – optimization criterion.

Finding the necessary composition factors by solving system (2.28) is usually expedient when it is possible to use linear equations.

When using quadratic equations $y_i = f(P_i)$ it is necessary to determine the possibility of including in the system linear equations $\partial y_i / \partial P_i = 0$ obtained by differentiating initial equations by the factors that are optimized. In some cases, other methods of solving the system (2.28) are possible [22].

The experimental studies of ash-containing mortars have allowed us to obtain quadratic mathematical models of all normalized properties. In this case, the design of ash-containing compositions can be performed in the following order. We determined the fly ash content and modifier additives close to optimal values. Analysis of the above experimental results allows us to recommend a fly ash content equal to 35...40% of the cement consumption for mortars with design compressive strength 10 MPa and higher and 60...70% for mortars of lower strength.

The consumption of superplasticizer SP-3 should be taken as 0.25...0.3% of the cement consumption for modifier PFM_1, 0.45...0.5% for modifier PFM_2. Higher values are taken when using very fine sand.

The optimal consumption of the air-entraining additive in the composition of the PFM_1 modifier is 0.04% of the cement consumption.

The consumption of Tylose as a water-retaining additive in the composition of PFM_2 should be taken as 0.2...0.25% of the cement consumption. A higher value is taken when using sand with module fineness $M_f > 1.5$. Calculation of the mortar composition performing in the following sequence.

1. We'll find the required W/C from regression equations that describe the normalized properties of mortars. In order, to find the W/C, mathematical models of compressive and bending strength (2.16–2.19), adhesion strength (2.12–2.15), and frost resistance (2.16, 2, 27) are used. If there are requirements for other properties of mortar mixtures or mortars, other regression equations are used. Of all the obtained W/C values, the smallest is chosen as one that satisfies all conditions.
2. Solving the model of mortar workability (2.8...2.9) with respect to X_4 and moving from coded to natural factors, we find the water consumption (W, kg/m^3) that provides the required cone penetration of mortar mixture.
3. Find the cement consumption (kg/m^3)

$$C = W:(W/C). \tag{2.29}$$

4. Based on the known cement consumption, we find the consumption of ash and additives – components of the multifunctional modifier according to the specified recommendations.
   The sand consumption is found from the equation:

$$\frac{C}{\rho_c} + \frac{\text{Ash}}{\rho_{ash}} + \frac{W}{\rho_w} + \frac{S}{\rho_s} + V_{\text{air}} = 1000, \tag{2.30}$$

   where, C, Ash, W, S_1 – volumes of cement, fly ash, water and sand; V_{air} – volume of air drawn into the mixture, l;
   ρ_c, ρ_{ash}, ρ_w, ρ_s –densities of cement, ash, water and sand in kg/l.
   From here:

$$S = 1000 - \left(\frac{C}{\rho_c} + \frac{\text{Ash}}{\rho_{\text{ash}}} + \frac{W}{\rho_w} + V_{\text{air}}\right)\rho_s \tag{2.31}$$

Let us calculate, for example, the composition of an ash-containing mortar mixture with the multifunctional modifier PFM$_1$. for mortar with compressive strength 10 MPa, F50 in terms of frost resistance, with an adhesion strength to the base of at least 0.2 MPa. We use fine sand with M_f=1.6 and density of 2.65 kg/l.

1. The ash consumption is taken as 40%, superplasticizer SP-3 – 0.25%, air-entraining additive – 0.04% of the cement mass. Then in the models of adhesive strength (2.12), compressive strength (2.16), and frost resistance (2.26)

$$X_2 = \frac{0,4 - 0,35}{0,35 - 0} = 0,14 \quad X_3 = \frac{0,25 - 0,35}{0,35 - 0} = -0,28,$$

$$X_4 = \frac{0,04 - 0,025}{0,025 - 0} = 0,6$$

2. Having solved equations (2.12), (2.16), and (2.26) with respect to X_1 and having passed from coded to natural factors, we obtain from the condition of compressive strength $(W/C)_1$=1.07; adhesion strength $(W/C)_2$=1.03; frost resistance $(W/C)_3$=1.25. Finally, we accept $(W/C) = (B/Ц)_2 = 1.03$.
3. From the workability model (2.8) with

$$X_1(I) = \frac{0,04 - 0,025}{0,025 - 0} = 0,6; \quad X_2 = \frac{0,25 - 0,35}{0,35 - 0} = -0,28;$$

$$X_3 = \frac{0,4 - 0,35}{0,35 - 0} = 0,14 \quad \text{and} \quad X_5 = \frac{1,03 - 0,8}{1,0 - 0,6} = 1,15 \text{ we get}$$

$$X_4 = -0,32, \text{ or } W = 270 + (-0,32) \cdot 30 = 260 \text{ kg/m}^3.$$

4. Cement consumption C = 260:1.03 = 252 kg/m^3.
5. Ash consumption Ash = 0, 4 C = 0, 4.252 = 101 kg/m^3.
   The consumption of air entraining additive AeA = 0, 0004.252 = 0.1 kg/m^3. Superplasticizer consumption SP = 0, 0025.252 = 0.63 kg/m^3.
6. The volume of drawn air V_{air} = 6% = 60 l/m^3.
   Then the sand consumption: S = 1476 kg/m^3.

The final composition of the mortar mixture: C = 252 kg/m^3, Ash = 101 kg/m^3, S = 1476 kg/m^3, AeA = 0.1 kg/m^3, SP-3 = 0.63 kg/m^3.

This composition is close to optimal from technological considerations, but does not take into account the cost of using the materials. The problem of choosing the optimal composition of the mixture for 3D printing taking into account the cost of materials, can be strictly solved by mathematical programming methods. Various optimality conditions are possible: minimizing the cost of the mortar mixture, minimizing cement consumption without or with restrictions on the consumption of one or two additive components, etc. In all cases, the mandatory restrictions are to ensure the required set of normalized properties of the mortar.

When minimizing the cost of mortar mixture (Cost_m), the optimization criterion can be expressed by a linear equation:

$$\text{Cost}_m = Pr_C C + Pr_{ad} Ad + Pr_S S \tag{2.32}$$

where, Pr_c, Pr_{ad}, Pr_s – price of cement, additives, and sand, respectively, C, Ad, and S – their consumptions.

For the design of the composition with the adopted optimality criterion, the method of successive approximations can be used. At the first stage, the possibility of ensuring normalized properties with a minimum content of additives is checked, then with a certain step in the content of additives, the calculation is repeated. Ultimately, such calculations, performed using a computer, allows one to choose the optimal composition.

Before performing calculations to determine the composition of the mixture, it is advisable to perform an analysis of mathematical models by the graphical or dissociative-step method [22]. According to the latter, for factors with quadratic coefficients of models $b_{ii} < 0$, the extremum, regardless of the experimental zone, is determined from the equality of zero of the first partial derivative of the function:

$$y = b_0 + \sum_{i=1}^{K} b_i X_i + \sum_{i=1}^{K} b_{ii} X_i^2 \sum_{i \neq h} b_{ij} X_i X_j;$$

$$\frac{dy}{dX_i} = \left(b_i + \sum_{i \neq j}^{K} b_{ji} X_j \right) + 2b_{ii} X_i; \tag{2.33}$$

The location of the extremum X_{ie} is determined by the linear dependence:

$$X_{ie} = \frac{b_i + \sum b_{ij} X_j}{2b_{ii}} = -\frac{b_i}{2b_{ii}} - \frac{\sum b_{ii} X_j}{2b_{ii}} \tag{2.34}$$

In order for X_{ie} to always be in the experimental zone ($X_{ie} \leq 1$), the following conditions must be met:

$$\left(|b_i| + \sum_{i \neq j} |b_{ij}| \right) \leq 2|b_{ii}|. \tag{2.35}$$

In the case where inequality occurs:

$$\left(|b_i| + \sum_{i \neq j} |b_{ij}| \right) \leq 2|b_{ii}|, \tag{2.36}$$

then additional conditions need to be checked:

$$\left(|b_i| + \sum_{i \neq j} |b_{ij}| \right) \geq 2|b_{ii}|; \tag{2.37}$$

$$\left(|b_i| + \sum_{i \neq j} |b_{ij}| \right) \leq 2|b_{ii}|. \tag{2.38}$$

In the case of fulfillment of condition (2.36), $X_{i.opt}$ is always outside the experimental zone and is equal to the limiting value ±1, which is determined by the sign b_i.

According to the proposed method, the compositions of masonry ash-containing mixtures for 3D mortars of grades 50...125 were calculated for the ratio of prices for starting materials. Table 2.16 shows the compositions of mortar mixtures using multifunctional modifiers.

Table 2.16 Compositions of masonry mortars for 3D printing based on modified mixtures.

Mortar Brand	*Consumption of Components per 1 m³ of Mortar, kg*							*N = S/ Binder*
	Cement	*Ash*	*SP-3*	*AeA*	*CE*	*ater*	*Sand*	
150	340	136	0,850	0,136	-	255	1374	2,9
	320	128	1,600	-	0,8	250	1468	3,3
	305	122	1,220	-	-	232	1587	3,4
100	255	102	0,638	0,102	-	255	1485	4,2
	240	96	1,200	-	0,6	247	1578	4,7
	230	92	0,920	-	-	242	1659	4,7
75	215	129	0,538	0,086	-	247	1510	4,4
	190	114	0,950	-	0,48	224	1662	5,5
	180	108	0,720	-	-	212	1761	5,6
50	150	105	0,375	0,06	-	225	1651	6,5
	140	98	0,700	-	0,35	217	1741	7,3
	135	94,5	0,540	-	-	204	1837	7,2

Calculations related to the design of modified construction mixtures can also be performed using nomograms obtained by graphically interpreting mathematical models (Fig. 2.22).

Dry construction mixes (DCM) of 3D printing can be recommended for production in two ways. The most appropriate is to produce a mixture in a factory that is completely ready for use, i.e. together with sand. In a simpler version, it is possible to produce a mixture without sand. In this case, in the factory, DCM is produced, which contains cement, ash, and modifier additives, and in the conditions of the construction site, this DCM is mixed in the proportions indicated in Table 2.16. However, in the latter case, the properties of the mortar will be influenced by the characteristics of the sand used, which may lead to some instability of the operational properties.

Using the example of mortars with the lowest modifier PFM-1 consumption for the same brand, the influence of the method of preparation of the dry mixture, as well as storage conditions on the

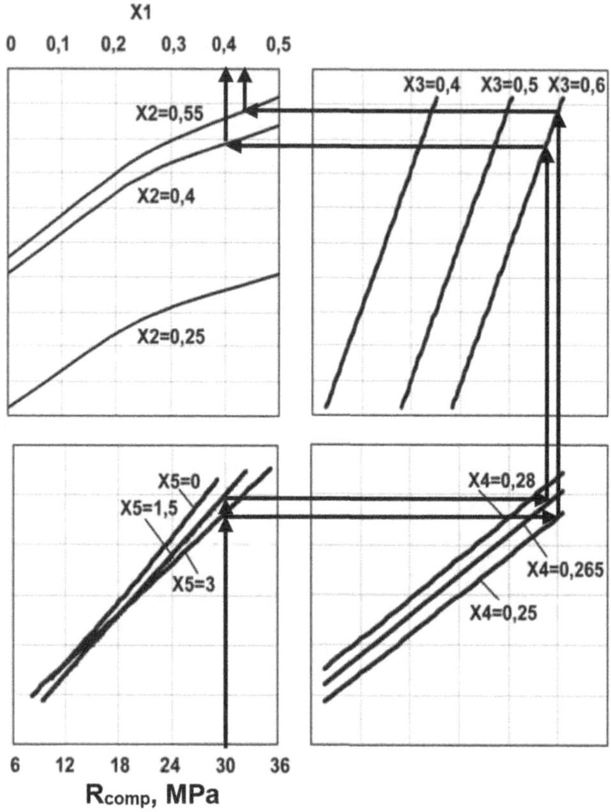

Fig. 2.22 Nomogram of compressive strength of mortars with PFM containing fly ash.

compressive strength of the mortar at the age of 28 days was studied. The water consumption was taken such that the penetration of the cone of the mortar mixture was 8...10 cm.

The PFM-1 additive was introduced with one-, two- and three-stage mixing (Table 2.17). In the first method, pre-mixed additives included in the modifier were introduced into the forced-action mixer simultaneously with other components - cement, ash and sand. In the second method, the modifier additives were pre-mixed with ash, which was then mixed with cement and sand. The third method consisted of mixing pre-prepared ash with the PFM_1 additive and a cement-sand mixture. The total mixing time varied from 3 to 15 minutes. The composition of the mixture was next: $C = 255$ кг/м^3, Ash/C = 0, 4; SP-3 – 0, 25% from cement consumption; AeA – 0, 04% from cement consumption; n = Sand/cementash binder = 4, 2 (Sand/Cement = 5, 9).

With the simultaneous introduction of all components of the dry mixture, the required homogeneity is achieved after 10 minutes of mixing, as evidenced by the stabilization of the achieved strength and, accordingly, the water consumption of the mortar mixture.

When pre-preparing ash with the addition of PFM, the duration of mixing the dry mixture can be reduced to 5 minutes, and when using a homogeneous ready-made cement-sand mixture, to 3 minutes. The implementation of two- and three-stage technologies for producing dry mixes is possible when using additional mixers designed for mixing ash with the addition of PFM, cement, and sand, and the finished mixture. Therefore, when choosing a technology option, the economic factor should be taken into account. Most likely, multi-stage technology is justified for large production volumes, when reducing the mixing time is highly desirable and justifies the use of additional equipment.

Table 2.17 The influence of the method of introducing PFM1 and the duration of mixing DCM on the compressive strength of masonry mortar.

No. of the Experiment	*Method #1*		*Method #2*			*Method #3*		
	Duration of Mixing DCM, min	*Mortar Strength, MPa/ Water Consumption kg/m³*	*Mixing Duration, min*		*Mortar Strength, MPa/ Water Consumption kg/m³*	*Mixing Duration, min*		*Mortar Strength, MPa/ Water Consumption kg/m³*
			PFM with Ash	*Dry Mix*		*PFM with Ash*	*Dry Mix*	
1	3	9,5/264	3	3	9,6/256	3	3	10,5/253
2	5	9,7/258	3	5	10,4/253	3	5	10,6/251
3	7	10,1/255	3	7	10,6/251	3	7	10,5/251
4	10	10,5/252	5	3	10,0/255	5	3	10,6/250
5	15	10,6/252	5	5	10,7/252			-
6			5	7	10,6/252			-

Note: In method #3, the cement-sand mixture was mixed for 3 minutes.

The effectiveness of introducing PFM_2 components in the process of ash and cement mixing was also investigated. The results of the experiments are given in Table 2.18. The composition of the mixture is similar, the mixing duration was 5 min. Analysis of the data in Table 2.18 shows that by introducing PFM into a ball mill during the grinding of ash or cement, it is possible to additionally increase the activity of the mixture by 10...20% due to some increase in the fineness of grinding of dispersed materials and uniform distribution of surfactant additives in them. The highest effect is achieved when ash and cement are mixed together with the addition of PFM.

In the practical use of dry building mixtures for 3D printing, their important technological characteristics are fluidity and stability of activity over time.

Table 2.18 The effect of the additional grinding during production of ash and cement mixture with PFM additives on the activity of the dry mix.

Obtaining Method	*Water Consumption kg/m^3*	*Strength of the Mortar, MPa*	*Grinding Duration, min*	*Specific Surface Area, m^2/kg*
Grinding of fly ash in a ball mill with the addition of PFM$_2$	252 246	11,1 11,5	30 60	360 385
Cement grinding with the addition of SP-3 and the introduction of CE into the mixer	250 260	11,2 11,9	30 60	350 378
Grinding of fly ash with the addition of SP-3 and introduction of CE into the mixer	254 250	10,9 11,4	30 60	355 384
Joint grinding of ash and cement with the addition of PFM	250 252	11,6 12,4	30 60	345 375

The fluidity of dry mixtures and their tendency to form lumps, as well as to hand in bunkers and other containers, depends on their ability to hygroscopically moisten, which was determined by the increase in the mass of the samples while maintaining $-\Delta m$ (Fig. 2.23). Cement-ash mixtures without PFM additives for up to 90 days at a humidity of 90% adsorb approximately 11% of moisture (relative to the mass of cement-ash binder), with the addition of PFM$_2$ – 9%, and with the addition of PFM$_1$ – 8%. The content of lumps larger than 1 mm in mixtures without PFM was about 50%, with PFM – 15–30%. At a relative humidity of 60%, the mixture without PFM adsorbed 7.5% of water, with PFM$_2$ – 6%, with PFM$_1$ – 5%. The number of lumps in the mixture was 10–20%.

The decrease in the sorption capacity of cements and building mixtures with surfactant additives has been repeatedly noted by researchers [41]. It is associated with the formation of surfactant adsorption layers on binder grains. The changing the type of PFM did not cause a sharp change in the sorption capacity of the cement-ash binder in dry mixtures of the studied

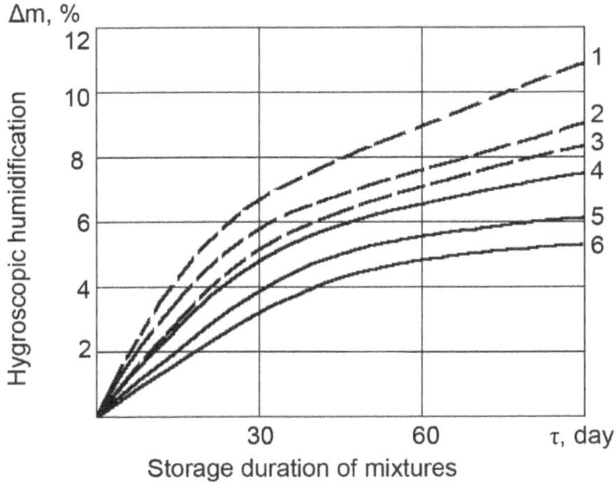

Fig. 2.23 The effect of the storage duration of dry mixtures on their hygroscopic moistening: – humidity 90%; – humidity 60%: 1 – without PFM; 2 – PFM_2; 3 – PFM_1; 4 – without PFM; 5 – PFM_2; 6 – PFM_1.

composition. Sorption moistening of the cement-ash binder during open storage of dry mixtures leads to a decrease in their activity, which is less typical in mixtures with PFM additives (Table 2.19).

Table 2.19 The effect of the duration of storage of dry mixes on the change in their activity.

No. of the Experiment	*Humidity, %*	*PFM Type*	*Compressive Strength of the Mortar, MPa, with Storage Time of the Mixture, Days*			
			1	*30*	*60*	*90*
1	60	-	9,4	8,4	7,5	5,9
2	90	-	9,7	8,0	6,9	5,1
3	60	PFM_1	10,6	9,7	8,9	8,2
4	90	PFM_1	10,8	9,4	8,6	8,0
5	60	PFM_2	10,7	10,1	8,8	8,0
6	90	PFM_2	11,2	9,7	8,4	7,9

3

Mixtures Based on Cement-Slag Binder with Modifying Additives

3.1 Mixtures Based on Cement-slag Binder with Hardening Accelerator Additive

The initial components for the studied mixtures were Portland cement CEM I 42,5N of "Volyn-cement", granulated blast furnace slag (GBS) of the Kryvyi Rih Metallurgical Plant (Ukraine). The chemical composition of the slag is as follows (%): SiO_2 – 39.51; Al_2O_3 – 6.47; Fe_2O – 30.14; CaO – 47.19; MgO – 3.12; SO_3 – 1.76; MnO – 1.14; LOI – 0.59. The basicity modulus is M_b – 1.27, the activity modulus is M_a – 0.16.

The share of glass phase in GBS was 75...80%. According to the basicity modulus, used slag can be attributed to basic slags in accordance with this indicator, this slag meets the requirements for slag as a component of cement-slag binders. In general, the values of M_b, M_a and the content of glass phase indicate high hydraulic activity of the slag.

The research studied the influence of the material composition of the mixtures and the amount of the hardening accelerator Na_2SO_4 additive to ensure the complex of necessary properties of the mixtures for 3D concreting. Cement-slag binder with the Na_2SO_4 additive was mixed with quartz sand (M_f= 1.9). To select the optimal parameters of the mixture composition, algorithmized experiments were conducted according to the typical three-level three-factor plan B_3 [22]. The conditions for planning the experiments are given in Table 3.1.

The water consumption was varied to achieve the required workability by penetration of the standard cone 8…10 cm, which ensures the extrudability of the mixture. Testing of the properties of mortar mixtures was carried out in accordance with DSTU B V.2.7-176:2009 (Ukrainian standard).

DOI: 10.1201/9781003685883-3

The planning matrix and experimental results determine structural strength (R_s) as well as tensile strength in splitting (R_{spl}) and compressive strength (R_{comp}) are given in Tables 3.2 and 3.3.

Table 3.1 Conditions for planning experiments.

Factors of Influence		*Levels of Variation*			*Variation Interval*
Natural View	*Coded View*	*–1*	*0*	*+1*	
The content of GBS by weight of the binder, %,	X_1	50	40	30	–10
Content of hardening accelerator additive, % of the binder mass. (HA)	X_2	0	1	2	1
Binder content, kg/m^3, (Bind.)	X_3	300	400	500	100

Table 3.2 Planning matrix and composition of mixtures for 3D printers.

No	*Factor Values*						*Mixture Composition, kg/m^3*				*Water Consumption, l/m^3*
	Coded			*Natural*							
	X_1	X_2	X_3	*GBS, %*	*HA %*	*Bind. kg/m^3*	*Cem*	*GBS*	*HA*	*Sand*	
1	1	1	1	30	2	500	350	150	10	1573	230
2	1	1	– 1	30	2	300	210	90	6	1839	200
3	1	– 1	1	30	0	500	350	150	0	1586	235
4	1	– 1	–1	30	0	300	210	90	0	1834	208
5	– 1	1	1	50	2	500	250	250	10	1577	225
6	– 1	1	– 1	50	2	300	150	150	6	1860	190
7	– 1	– 1	1	50	0	500	250	250	0	1590	230
8	– 1	– 1	– 1	50	0	300	150	150	0	1850	200
9	1	0	0	30	1	400	280	120	4	1650	240
10	– 1	0	0	50	1	400	200	200	4	1696	220
11	0	1	0	40	2	400	240	160	8	1702	215
12	0	– 1	0	40	0	400	240	160	0	1697	225
13	0	0	1	40	1	500	300	200	5	1582	230
14	0	0	– 1	40	1	300	180	120	3	1797	218
15	0	0	0	40	1	400	240	160	4	1700	220
16	0	0	0	40	1	400	240	160	4	1700	220
17	0	0	0	40	1	400	240	160	4	1700	220

Table 3.3 Experimental results of studies of mixture properties.

No	*Duration of Holding (D_h), min*	*Structural Strength, Pa, after Mixing*			*Strength at the Age of 3 Days, MPa*		*Strength at the Age of 28 Days, MPa*	
		10 min	*25 min*	*40 min*	R^3_{spl}	R^3_{comp}	R^{28}_{spl}	R^{28}_{comp}
1	35	5320	8513	15160	6.6	19.7	9.3	32.7
2	60	3780	5679	10700	4.5	10.8	6.7	21.2
3	85	3916	5482	9225	5.7	16.9	8.1	27.6
4	145	2730	4096	7000	3.9	7.8	6.1	19.5
5	70	4600	6916	12300	4.9	13.8	6.6	19.4
6	95	3356	5034	9125	4.1	8.8	5	13.4
7	120	3426	5139	7480	4.4	13.1	6.1	17.1
8	160	2521	3781	6545	3.5	7.8	4.8	9.4
9	55	3637	5455	9375	4.3	12.9	7.7	21.7
10	85	3177	4765	7710	3.6	10,6	5.1	16.4
11	50	4020	6030	11230	4.7	12	6.5	20.5
12	120	2900	4350	6970	4.0	12.7	6.5	18.5
13	50	4002	6003	9880	6.3	17.8	8.7	28.5
14	95	2782	4173	7180	4.2	9.4	5.2	13.0
15	65	3275	4913	8240	5.0	10.9	7.2	21.7
16	70	3300	4950	8260	4.7	10.6	6.9	21.4
17	65	3310	4965	8200	4.9	10.8	7.1	21.6

After processing and statistical analysis of experimental data, regression equations for the properties of concrete mixtures and concretes were obtained, which are given in Table 3.4.

Based on the results obtained, graphical dependences (Figs. 3.1…3.6) of the properties of concrete mixtures and concretes for a 3D construction printer using GBS and a hardening accelerator additive were constructed.

Table 3.4 Regression equation of properties of concrete mixtures and concretes for 3D construction printer.

Initial Parameters	*Statistical Models*	*Regression Equations*
Duration of holding (D_h), min	$D_h = 66{,}4 - 15{,}0 \cdot x_1 - 32{,}0 \cdot x_2 - 19{,}5 \cdot x_3 + 4{,}1 \cdot x_1^2 + 19{,}1 \cdot x_2^2 + 6{,}6 \cdot x_3^2 - 2{,}5 \cdot x_1x_2 - 2{,}5 \cdot x_1x_3 + 6{,}3 \cdot x_2x_3$	(3.1)

Continuation of Table 3.4

Initial Parameters	*Statistical Models*	*Regression Equations*
Structural strength, Pa, 10 min after mixing, Pa	$R_S^{10\,\min} = 3295 + 230 \cdot x_1 + 558 \cdot x_2 + 610 \cdot x_3 + 124 \cdot x_1^2 + +177 \cdot + \cdot x_2^2 + 108 \cdot x_3^2 + 55 \cdot x_1x_2 + 72 \cdot x_1x_3 + 87 \cdot x_2x_3$	(3.2)
Structural strength, Pa, 40 min after mixing, Pa	$R_S^{40\,\min} = 8258 + 830 \cdot x_1 + 2130 \cdot x_2 + 1350 \cdot x_3 + 296 \cdot x_1^2 + 853 \cdot x_2^2 + 283 \cdot x_3^2 + 280 \cdot x_1x_2 + 322 \cdot x_1x_3 + 560 \cdot x_2x_3$	(3.3)
Compressive strength at the age of 3 days, MPa	$R_{comp}^3 = 11{,}66 + 1{,}4 \cdot x_1 + 0{,}68 \cdot x_2 + 3{,}67 \cdot x_3 - 0{,}537 \cdot x_1^2 + 0{,}063 \cdot x_2^2 + 1{,}313 \cdot x_3^2 + 0{,}513 \cdot x_1x_2 + 0{,}963 \cdot x_1x_3 - 0{,}062 \cdot x_2x_3$	(3.4)
Tensile strength at splitting at the age of 28 days, MPa	$R_{spl}^{28} = 6{,}89 + 1{,}03 \cdot x_1 + 0{,}25 \cdot x_2 + 1{,}1 \cdot x_3 - 0{,}328 \cdot x_1^2 - 0{,}228 \cdot x_2^2 + 0{,}222 \cdot x_3^2 + 0{,}138 \cdot x_1x_2 + 0{,}213 \cdot x_1x_3 + +0{,}113 \cdot x_2x_3$	(3.5)
Compressive strength at 28 days, MPa	$R_{comp}^{28} = 20{,}74 + 4{,}7 \cdot x_1 + 1{,}51 \cdot x_2 + 4{,}88 \cdot x_3 - 1{,}01 \cdot x_1^2 - 0{,}56 \cdot x_2^2 + 0{,}69 \cdot x_3^2 + 0{,}06 \cdot x_1x_2 + 0{,}74 \cdot x_1x_3 + 0{,}21 \cdot x_2x_3$	(3.6)

The expiry date life of mixtures for 3D concreting based on blast furnace granulated slag is reduced from 2.5 h to 40…45 min at a binder consumption of 500 kg/m^3, a minimum proportion of GBS in the binder (30%) and the use of a hardening accelerator additive in the amount of 2% of the binder (Fig. 3.1…3.3). It should be noted that a more significant effect of the hardening accelerator is observed at a dosage of up to 1%, and with a further increase in its content, a positive character is observed only under the condition of a minimum proportion of GBS in the binder. A significant reduction in the expiry date is achieved by increasing the amount of cement-slag binder.

From Figs. 3.4 and 3.5 it is noticeable that the mortars show the lowest values of compressive strength at the age of 3 and 28 days in the absence of a hardening accelerator additive in the composition and a minimum binder content. Under the condition of introducing a hardening accelerator and a gradual decrease in the proportion of GBS, a more active increase in strength is observed with the consumption of the hardening accelerator additive. An increase of the binder content in the mixture

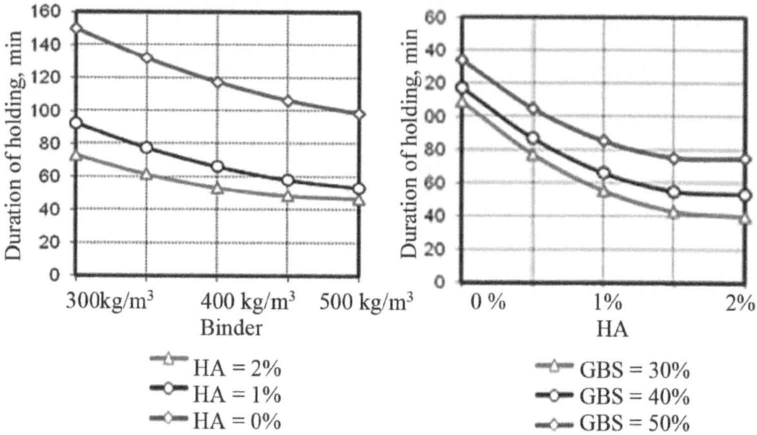

Fig. 3.1 Graphical dependences of the mortars duration of holding for a 3D printer using GBS and the additive of a hardening accelerator (HA).

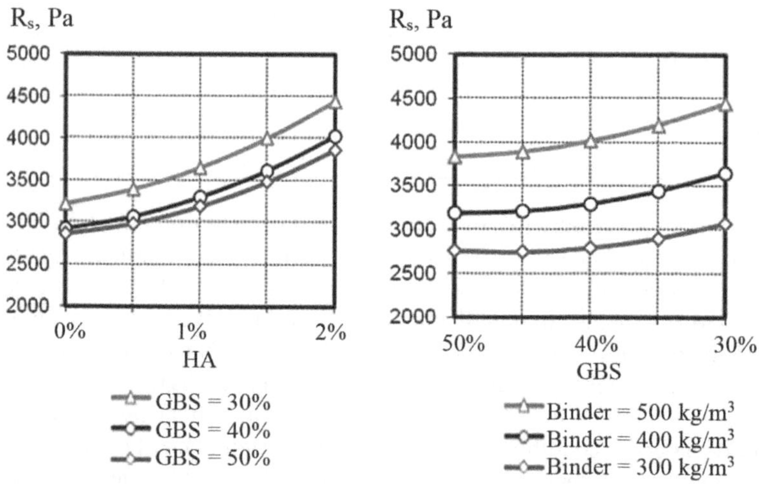

Fig. 3.2 Graphical dependences of the structural strength (Rs) of concretes for a 3D printer 10 min after mixing using GBS and a hardening accelerator additive (HA).

causes a much more rapid increase in strength, compared to an increase in the proportion of hardening accelerator, which has a linear nature of the effect.

According to the obtained graphical dependencies (Fig. 3.6), with a decrease in the proportion of GBS from 50% to 40% of the binder mass

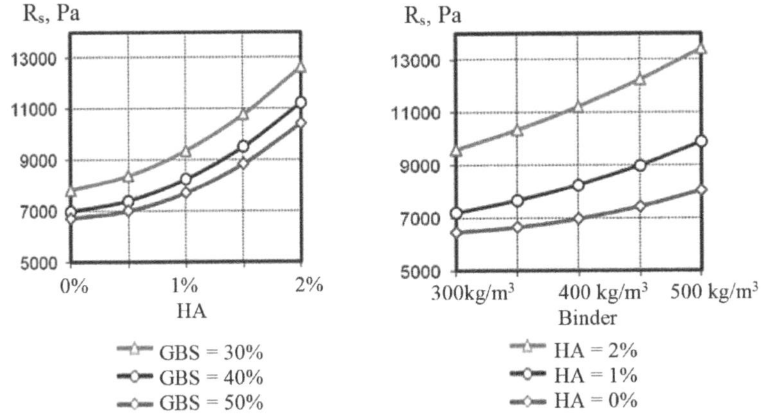

Fig. 3.3 Graphical dependences of the structural strength of mortars 40 min after mixing using GBS and the addition of a hardening accelerator.

and a gradual increase of the binder from 300 kg/m^3 to 500 kg/m^3, an increase in the 28-days compressive strength of concretes is observed by 35...40% with a GBS content of 40...50% and by 15...20% with a minimum proportion of slag. From the above graphical dependencies, it is noticeable that the effect of the addition of the hardening accelerator is also positive, which is more pronounced at the maximum binder content. In this case, the effective accelerator dosage is the content of 0.7...1.3% of binder mass; a further increase does not lead to a significant increase in strength.

3.2 Clinker Kiln Dust Removal: An Active Component of Cement-slag Binder Systems

Clinker kiln includes particles of the raw material mixture for producing cement clinker with a size of less than 1 mm, which are carried out of the kilns and captured in dust collection devices. Dust emissions during kiln operation account for up to 80% of the total amount of dust generated in cement production [60, 61]. Depending on the size and type of kilns, the presence of heat exchange devices in them, the firing mode, the method of cement production, the properties of the raw material mixture and fuel, the dust emission on average ranges from 5 to 20% of the mass of the mixture fed into the kiln. It has been established that the greatest amount of dust is carried away by flue gases from the clinker kiln at the end of the chain curtain zone and also the calcination zone. This dust is formed mainly as a result of mechanical destruction by the chains and abrasion of raw

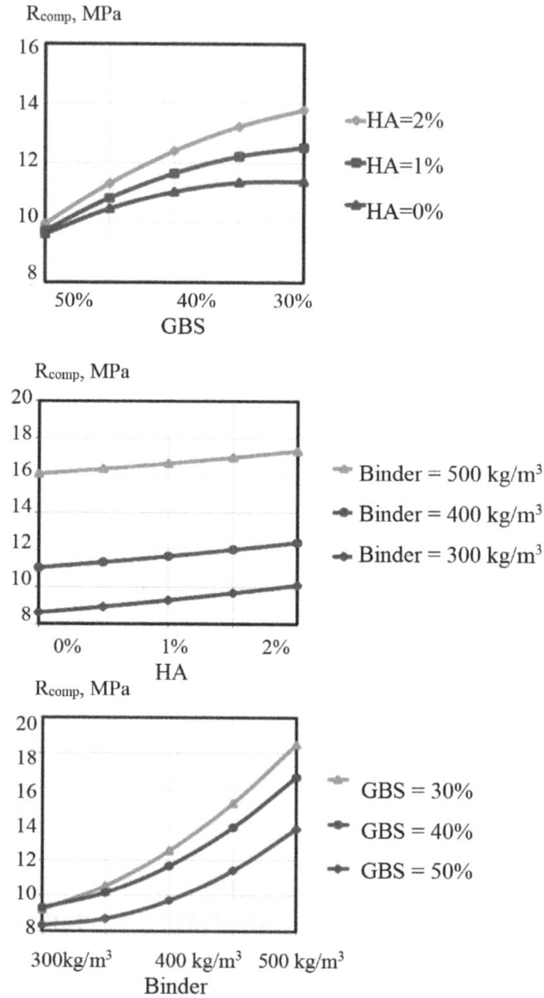

Fig. 3.4 Graphical dependences of compressive strength at the age of 3 days of the mortars using GBS and the addition of a hardening accelerator.

material granules on the furnace lining. Significant amounts of dust can also be generated in the zones of exothermic reactions and cooling. High dust emission is characteristic of the use of non-plastic raw materials, increased surface area of heat exchangers, and gas flow velocity.

In terms of material composition, the dust consists mainly of a mixture of undecomposed limestone and burnt clay [62]. Compared to the original raw material mixture, the dust is characterized by a higher

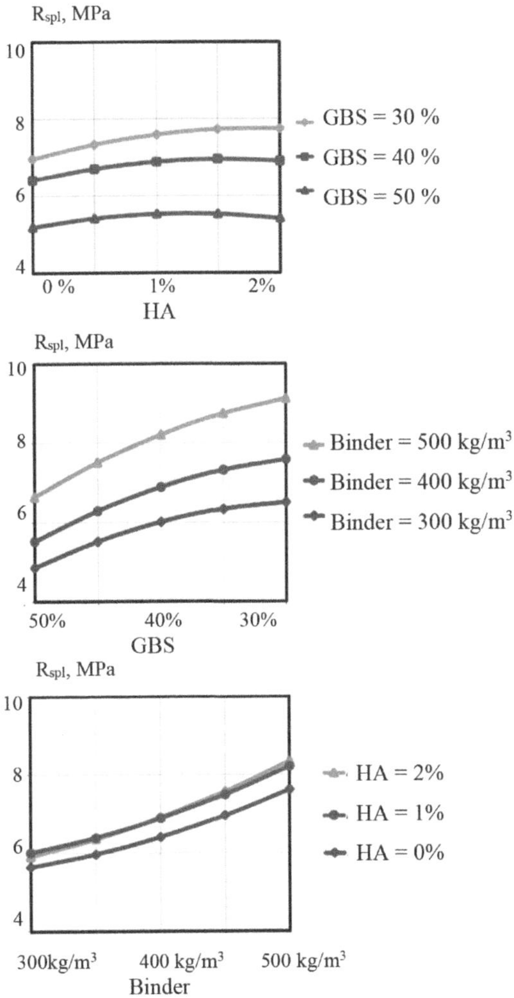

Fig. 3.5 Graphical dependences of tensile strength at splitting at the age of 28 days of the mortars using GBS and the addition of a hardening accelerator.

content of the carbonate component and a lower content of the clay and ferruginous components. The dust may contain 10–20% clinker minerals (β and γC_2S – 8–10%, C_2F and C_4AF – 10–12%). The content of free CaO increases with an increase in the content of clinker minerals in the dust from 2 to 14%.

One of the characteristic features of dust is its high content of alkaline compounds, represented by sulfates, carbonates, and bicarbonates of

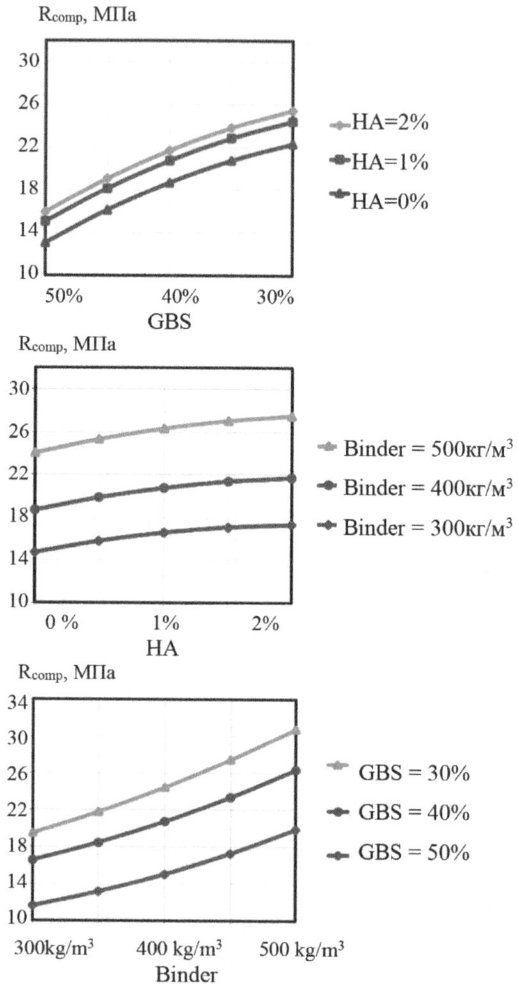

Fig. 3.6 Graphical dependences of the compressive strength of mortars at the age of 28 days using GBS and the addition of a hardening accelerator.

sodium and potassium, as well as silicates of variable composition. The alkali content in dust usually does not exceed 8% [63], although data are given [64] that the total amount of $K_2O + Na_2O$ in dust collected by electrostatic precipitators can vary from 0.3 to 45% with a predominance of potassium oxide. The alkali content increases in the finest fractions. In dust accumulated in the third field of the electrostatic precipitator, it can be 3–4 times more than in the bins of the I and II fields.

The higher alkali content in the dust is explained by the peculiarities of the chemical processes occurring in rotary kilns during clinker burning [65]. With increasing temperature in the clinker kiln, alkalis, which are mainly present in the raw material mixture and fuel ash in the form of silicates, are displaced by calcium oxide.

The main method of clinker kiln dust disposal is its return to the clinker kiln production process. Various methods of dust return have been developed [66]: mixing it with sludge or raw flour entering the kiln in the form of dry powder, granules, or sludge; feeding it through chain curtains in a dusty or granular state; blowing it into the kiln from the hot end. Methods have also been developed for firing the dust after adjusting its composition in a separate kiln and pre-leaching it before returning it to the kiln. Returning the dust by any means complicates the process of firing cement clinker and usually negatively affects the quality of the cement due to the increase in its alkali content, increased heterogeneity of the chemical composition, and deterioration of the clinker structure.

Along with the return of the dust to the raw material mixture for clinker production, a number of other areas of its application have been developed, including for the production of binders and mortars [67–69].

Dust formed during the firing of Portland cement clinker and captured by electrostatic precipitators has weak binding properties. They are due to the presence in it of a certain amount of clinker minerals formed as a result of mainly solid-phase reactions. A certain contribution to the hydraulic activity of dust can be made by the interaction of free lime and dehydrated clay contained in it. Due to this activity, as well as high dispersion, which contributes to easy workability of mortar mixtures and prevention of water separation, the dust is recommended as an additive to building mortars.

The hydraulic activity of dust increases significantly with the addition of gypsum. According to V.S. Batalin [69], the addition of gypsum in an amount of 2...3% allows, already in the first stages of hardening, to increase the strength of samples by more than three times compared to samples made from pure dust. At the same time, the increased gypsum content negatively affects the final strength of the binder.

The best results were obtained in the compositions of dust-blast furnace granulated slag and dust-slag-gypsum [70]. Along with blast furnace slag, the possibility of using phosphoric and cupronickel slags in combination with dust was also noted. The theoretical prerequisite for the compatibility of slag and cement dust in the production of composite binder is mainly the presence in the latter of soluble alkalis, which have the strongest activating effect on slag glass. Dust components such as free oxide and calcium sulfate also have a certain activating effect on granulated slag.

In the soil silicate laboratory of the Kyiv Civil Engineering Institute, a slag-alkaline binder based on granulated blast furnace slag and high-alkaline dust from electrostatic precipitators of clinker kilns was first obtained [70]. When determining the optimal ratio between slag and dust, it was found that the highest performance was achieved with a binder composition of 75% slag and 25% dust. When mixed with water, the binder grade varied from 20 to 40 in MPa depending on the amount of alkaline carbonates in the dust, and when mixed with alkaline solutions, from 30 to 70, depending on the concentration of the alkaline solution.

V.D. Glukhovsky and N.L. Makedon established [71] a number of features of dust-slag binders – the increase in activity is proportional to the concentration of alkalis in the dust and increased bending strength. Based on these binders, it is possible to obtain mortars and concretes with a wide range of grades in terms of strength, frost resistance, water resistance, which have high adhesion to reinforcement and reliably protect it from corrosion. The binders obtained by activating the blast furnace or phosphoric slags with cement dust, are recommended for the construction of road bases from stone materials and soils.

It is recommended to divide dust into three grades depending on the indicators of chemical and hydraulic activity. The chemical indicator of dust activity is proposed to be calculated as the ratio of the total amount of CaO in dust (CaO_{total}) excluding CaO bound in gypsum to CaO and bound in dicalcium silicate, tricalcium aluminate, and tetracalcium aluminoferrite.

$$M = \frac{CaCo_{total} - 1{,}2L.O.I. - 0{,}7(SO_3 - 0{,}85R_2O)}{2{,}8SiO_2 + 1{,}65A/_2O_3 + 0{,}7Fe_2O3}, \qquad (3.7)$$

where, L.O.I. – losses upon calcination;
R_2O – alkaline oxide content ($Na_2O + K_2O$).

The hydraulic index of dust activity (R) is determined by the following formula:

$$R = R_{aver} M, \qquad (3.8)$$

where R_{aver} – average 28 days cement strength (R_{aver} = 40МПа).

For the first grade dust, the value of the modulus M should be greater than 0.25, $R > 10$, for the second grade – $M = 0.125...0.25$, $R = 5...10$; for the third grade $M < 0.125$, $R < 5$.

The first grade dust can be used as an independent binder, the second and third grade as a component of the dust-slag binder, respectively, in the amount of 15...25 and 20...30%.

3.3 Rheological and Structural-mechanical Properties of Aqueous Pastes in Dust-slag-superplasticizer and Dust-slag-cement-superplasticizer Systems

Normal consistency and setting time. At the first stage, the change in normal consistency and setting time of pastes in the dust-slag-superplasticizer and dust-slag-cement-superplasticizer systems was investigated. Composite binders were obtained in two ways:

1. mixing ground granulated blast furnace slag (S_{spec} = 320 m^2/kg), fly ash and superplasticizer SP-1 until a homogeneous mixture;
2. joint grinding of the specified components to a certain specific surface area.

Taking into account the inevitable fluctuations in the composition of dust, caused by fluctuations in the composition of the raw material mixture and the operating parameters of clinker firing, two dust samples were selected for research, the chemical and grain composition of which, as well as the specific surface area, are given in Tables 3.5 and 3.6.

Table 3.5 Chemical composition of dust.

Dust Sample	*Composition of Oxides, %*								*L.O.I., %*
	CaO	CaO_{free}	SiO_2	Al_2O_3	Fe_2O_3	MgO	SO_3	R_2O	
D_1	42,5	3,1	19,6	5,3	4,1	0,6	2,4	2,9	19,1
D_2	41,4	4,2	18,5	4,4	3,9	0,7	3,3	4,9	19,5

Table 3.6 Specific surface area and grain composition of dust.

Dust Sample	*Specific Surface Area, m^2/kg*	*Number of Fractions in % by Mass with Particle Size in Microns*				
		0–20	*20–40*	*40–60*	*60–90*	*90–200*
D_1	396	51,5	25,4	9,3	11,5	2,1
D_2	410	53,3	23,6	10,2	10,3	2,5

The normal consistency and setting times of the initial components of composite binders are given in Table 3.7. The results of determining these parameters for binders obtained by mixing the initial components are given in Table 3.8, and for joint grinding in Table 3.9.

Table 3.7 Normal consistency and setting times of the starting components of composite binders.

Number in Order	*Type of Output Components*	*Normal Consistency, %*	*Setting Time Hours-minutes*	
			Initial	*Final*
1	Portland cement	25,4	2…05	4…10
2	Blast furnace slag	29,2	9…10	15…20
3	Dust (D_1)	40,5	4…30	9…40
4	Dust (D_2)	42,4	1…30	8…20

The basic granulated slag used in the work, when ground to a specific surface area comparable to the specific surface area of Portland cement, is capable of hardening, although much slower than Portland cement (Table 3.7).

An increase in the content of alkaline compounds in dust causes a significant acceleration of the setting initially.

Table 3.8 Normal consistency and setting times of pastes based on composite binders obtained by mixing the starting components.

No.	*Binder Composition (ratio of components by weight)*	*Content SP, % by Mass*	*Normal Consistency, %*	*Setting Time Hours-minutes*	
				Initial	*Final*
1	Slag:dust (D_1)-1:1	-	35	8…18	14…10
2	Slag:dust (D_2)-1:1	-	33,4	6…15	12…20
3	Slag:dust (D_1)-1:1	1	28,5	6…25	11…50
4	Slag:dust (D_2)-1:1	1	28,2	5…15	10…40
5	Cement:slag:dust (D_1)-2:1:1	-	31,5	3…30	6…50
6	Cement:slag:dust (D_2)-2:1:1	-	27,4	2…40	5…30
7	Cement:slag:dust (D_1)-2:1:1	1	25,5	2…30	5…10
8	Cement:slag:dust (D_2)-2:1:1	1	24,8	2…10	4…30

The combination of slag and dust allows to significantly accelerate the setting time of compositions based on them, which indicates a significant intensification of the processes of structure formation when these two substances are combined. This effect can be explained by the combined activation of slag by alkalis, sulfates, and free lime contained in the dust.

Some effect accelerating the setting is observed (Tables 3.8, 3.9) when introducing a superplasticizer into the binder composition and increasing the fineness of grinding. In the first case, the setting rate is obviously affected by a decrease in the water-binding ratio, in the second, by an increase in the reactivity of the binder, which contributes to the faster formation of the coagulation structure and crystal nuclei of hydrated neoplasms.

Table 3.9 Normal consistency and setting time of pastes based on composite binders obtained by co-grinding the starting components.

No.	*Type of Binder*	*Content of SP, % by Mass*	*Specific Surface Area, m^2/kg*	*Normal Consis-tency, %*	*Setting Time Hours-minutes*	
					Initial	*Final*
1	Slag+dust (D_1)	-	350	29,2	2…10	8…30
2	Slag+dust (D_2)	1	370	23,1	1…40	6…40
3	Slag+dust (D_1)	-	580*	31,4	1…30	4…50
4	Slag+dust (D_2)	1	660*	21,8	0…40	1…30
5	Slag+dust (D_1)	-	340	28,4	1…40	6…30
6	Slag+dust (D_2)	1	380	22,6	1…10	5…50
7	Slag+dust (D_2)	-	590*	31,5	1…10	3…50
8	Slag+dust (D_2)	1	640*	22,9	0…10	1…10
9	Cement+slag+dust (D_1)	-	330	27,4	0…30	5…10
10	Cement+slag+dust (D_1)	1	360	22,3	0…38	4…30
11	Cement+slag+dust (D_1)	-	590*	29,5	0…25	1…10
12	Cement+slag+dust (D_1)	1	630*	22,5	0…21	0…40
13	Cement+slag+dust (D_2)	-	320	28,1	0…15	4…50
14	Cement+slag+dust (D_2)	1	350	22,7	0…20	3…39
15	Cement+slag+dust (D_2)	-	620*	31,4	0…17	0…45
16	Cement+slag+dust (D_2)	1	630*	23,1	0…15	0…37

*– a grinding intensifier was introduced – polyglycol (0.03% of the binder mass)

Note: The composition of the binders is slag + dust – 1:1 (by weight), cement + slag + dust – 2:1:1 (by weight).

The introduction of Portland cement into the binder based on blast furnace slag and clinker kiln dust during co-grinding sharp accelerates the setting times. For the three-component composite binder obtained during co-grinding, they become significantly shorter than for Portland cement (the initial setting reaches 15–20 minutes (Table 3.9). With conventional mixing of Portland cement with ground slag and dust, the setting time of the binder is also accelerated, although to a much lesser extent (Table 3.8). It is obvious that when the components are ground together, mechanical and chemical activation of the composite binder occurs.

Bringing the total amount of dihydrate gypsum in the binder composition to 7...8% (SO_3 content is 4...5% by mass) allows to prove the setting times of composite binders to those recommended by the standard for Portland cement.

The normal consistency of composite binders ranges from 35 to 22% depending on the composition, method of production and dispersion. Binders obtained by co-grinding the components have a slightly lower normal consistency than those obtained by mixing. In the process of joint grinding, primary mechanical and chemical processes may occur and the amount of water demand begins to significantly deviate from that found from the condition of additivity of the water demands of individual components.

The most noticeable dilution effect on both two- and three-component composite binders is provided by the addition of a superplasticizer. With a specific surface area of 320–370 m^2/kg, the addition of 1% SP-1 by weight of the binder reduces its normal consistency by 18–21% (Fig. 3.7). Increasing the fineness of grinding of binders to 580...660 m^2/kg leads to a slight increase in normal consistency, which is compensated by the introduction of a superplasticizer additive.

The viscosity of water pastes based on composite binders largely determines the workability of mortar mixtures. To study the influence of the composition of the pastes on their viscosity, the rotational rheometry method and the RV-8M viscometer were used. An important advantage of rotational rheometry is the possibility of realizing practically unlimited shear deformations with its help [57]. The viscosity of the pastes was investigated using algorithmized experiments performed according to the typical B_4 plan [22]. The viscosity of the pastes (η) was determined in 2...3 min after mixing and calculated by the formula:

$$\eta = k\frac{P - P_0}{N}\tau, \quad (3.9)$$

where, k – instrument constant;

P – weight of cargo;

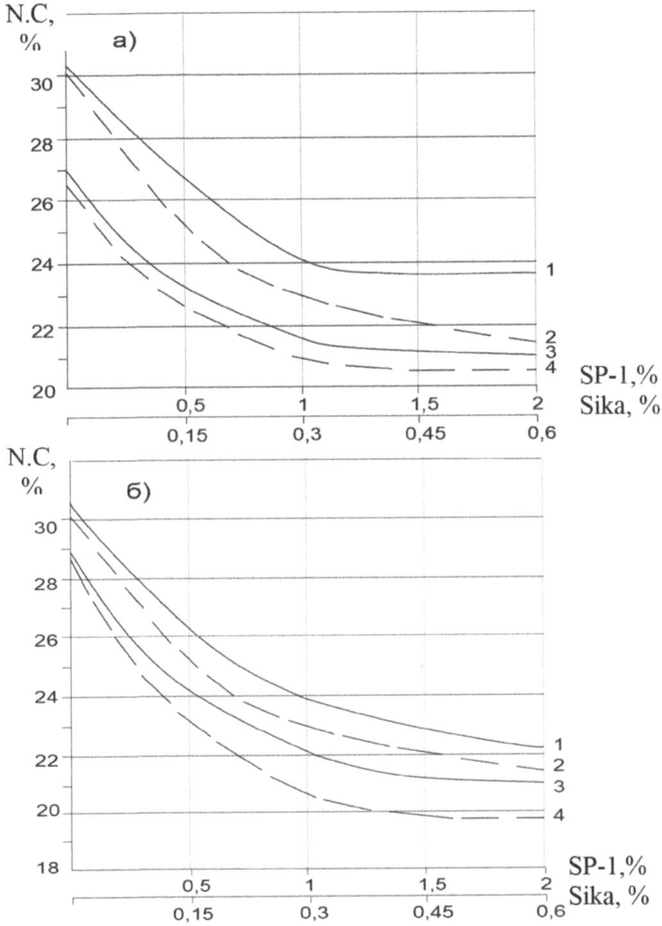

Fig. 3.7 The influence of the type and amount of superplasticizer additive on the normal consistency of composite binders obtained by joint grinding of the initial components: Binders with the addition of SP SP-1%; - - binders with SP Sika additive. (a) S = 330 – 370 m^2/kg; 1, 2 – slag:dust – 1:1; 3, 4 – Portland cement:slag:dust – 2:1:1. (б) S = 580 – 660 m^2/kg; 1, 2 – slag:dust – 1:1; 3, 4 – Portland cement:slag:dust – 2:1:1

N – number of revolutions;
τ – duration of one revolution;
P_0 – device idling load (P_0 = 12.4 g).

The conditions of planning experiments for determining viscosity are given in Table 3.10, the B_4 plan matrix and the results of the experiments are given in Table 3.11.

Table 3.10 Conditions for planning experiments when studying the viscosity of pastes.

No	*Factors*		*Levels of Variation*			*Variation Interval*
	Name	*Coded View*	*– 1*	*0*	*+1*	
1	Volumetric dust concentration in the dust-slag filler (DSF) – C_1	X_1	0	0,5	1	0,5
2	Volumetric concentration of DSF in composite cement-dust-slag binder – C_2	X_2	0	0,25	0,5	0,25
3	Volumetric concentration of composite binder in aqueous paste C_3	X_3	0,7	0,8	0,9	0,1
4	Content of superplasticizer (SP-1) in composite binder, % by mass – C_4	X_4	0	0,5	1	0,5

Table 3.11 Planning matrix and results of experiments on studying the viscosity of aqueous pastes based on composite binders.

Plan Points	*Factors*				*Viscosity, 10^{-1} Pa·s*	
	X_1	X_2	X_3	X_4	η_1	η_2
1	+1	+1	+1	+1	38,5	32,3
2	+1	+1	+1	– 1	53,2	49,4
3	+1	+1	– 1	+1	31,4	25,3
4	+1	+1	– 1	– 1	42,6	39,3
5	+1	– 1	+1	+1	19,7	18,7
6	+1	– 1	+1	–1	27,8	28,1
7	+1	– 1	– 1	+1	16,3	15,4
8	+1	– 1	– 1	– 1	25,6	26,1
9	– 1	+1	+1	+1	28,1	35,4
10	– 1	+1	+1	– 1	38,5	41,4
11	– 1	+1	– 1	+1	19,5	16,1
12	– 1	+1	– 1	– 1	27,8	27,5
13	– 1	– 1	+1	+1	23,5	24,3
14	– 1	– 1	+1	– 1	33,3	32,8
15	– 1	– 1	– 1	+1	15,8	15,5
16	– 1	– 1	– 1	– 1	24,8	25,5

Plan Points	*Factors*				*Viscosity, 10^{-1} Pa·s*	
	X_1	X_2	X_3	X_4	η_1	η_2
17	+1	0	+1	0	29,4	28,3
18	– 1	0	+1	0	22,3	21,6
19	0	+1	0	0	31,7	30,4
20	0	– 1	0	0	26,7	27,8
21	0	0	+1	0	30,8	31,5
22	0	0	–1	0	24,6	23,7
23	0	0	0	+1	21,2	21,6
24	0	0	0	–1	34,4	35,5

Note: η_1 – viscosity of the binder obtained by mixing the initial components; η_2 – viscosity of the binder obtained by joint grinding of the initial components to S = 350…380 m^2/kg.

Statistical processing of experimental data allowed to obtain the following regression equations, adequate at 95% confidence level, which can be considered as experimentally statistical models of the effective viscosity of aqueous pastes based on composite binders in the system of Portland cement-granulated blast furnace slag-dust of clinker kilns:

$$\eta_1 = 27{,}16 + 2{,}85X_1 + 5{,}48X_2 + 3{,}64X_3 - 5{,}26X_4 - 1{,}35X_1^2 + 1{,}99X_2^2 + 0{,}49X_3^2 + 0{,}59X_4^2 + 3{,}74X_1X_2 - 0{,}76X_1X_3 - 0{,}36X_1X_4 + 0{,}95X_2X_3 - 0{,}53X_2X_4 - 0{,}33X_3X_4 \quad (3.10)$$

$$\eta_2 = 27{,}32 + 1{,}28X_1 + 4{,}64X_2 + 4{,}45X_3 - 5{,}66X_4 - 2{,}41X_1^2 + 1{,}74X_2^2 + 0{,}24X_3^2 + 1{,}19X_4^2 + 2{,}23X_1X_2 - 1{,}68X_1X_3 - 0{,}96X_1X_4 + 1{,}81X_2X_3 - 0{,}62X_2X_4 + 0{,}32X_3X_4 \quad (3.11)$$

A graphical analysis of the main dependencies resulting from the obtained models is shown in Fig. 3.8. It can be seen from its that in the studied factor space the viscosity of pastes can vary almost 4 times, reaching the highest values at the maximum content of dust, minimum water content and the absence of a superplasticizer additive. With a decrease of the dust content in the slag-dust filler, its effect on the increase in viscosity decreases significantly. The content of superplasticizer in the paste based on composite binders is the strongest compensating factor that stabilizes the increase in viscosity when the pastes are saturated with dust.

The second factor in terms of the intensity of influence on the viscosity of pastes at a certain content of pulverized slag filler is the

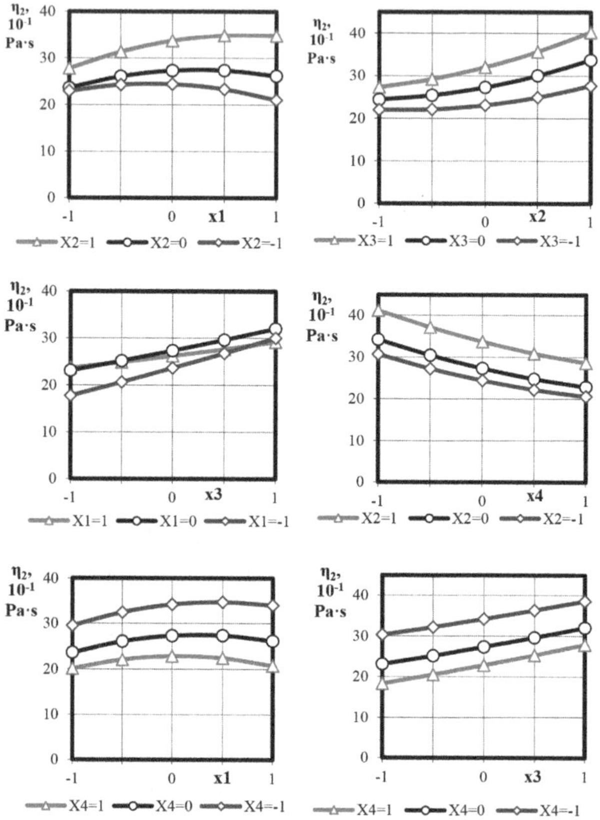

Fig. 3.8 Viscosity of water pastes based on composite binder obtained by joint grinding of the starting components to S = 350– 380 m^2/kg

Note: Apart from the two studied, the remaining variable factors are at the zero (base) level.

volumetric concentration of the composite binder in the aqueous paste, which directly reflects the influence of the water-binding ratio.

Regression equations (3.10, 3.11) allow, depending on the value of the factor X_3 and, accordingly, on the water-binding ratio, to regulate the dose of superplasticizer, which ensures a constant viscosity depending on the filler content and the ratio of slag and dust in it (Fig. 3.8). The nature of the influence of the studied factors on viscosity is quite similar for pastes obtained by mixing and co-grinding of binder components.

The features of the initial hardening of binder-water dispersions are characterized by the plastic strength, which is determined by the ultimate shear stress of the dispersed system in a certain period. According to

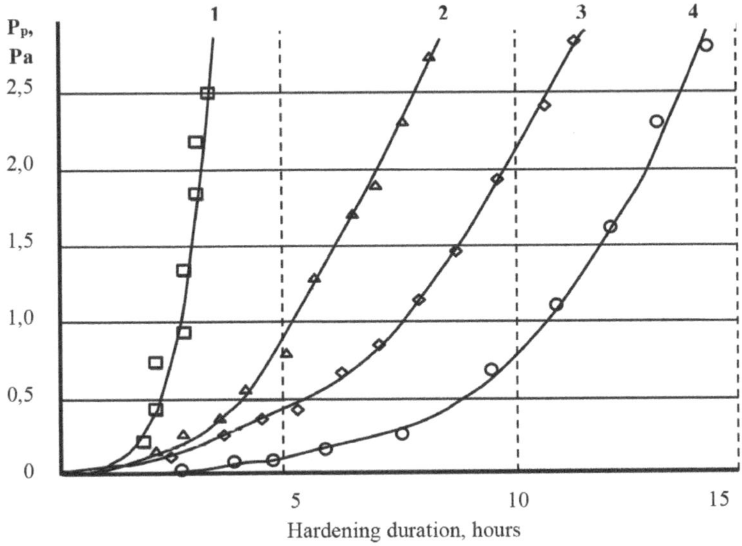

Fig. 3.9 Plastic strength of pastes based on two-component binders of slag-dust composition (D1) 1:1 (by mass): 1 – S = 660 m²/kg, 1% SP-1; 2 – S = 580 m²/kg; 3 – S = 370 m²/kg, 1% СП-1; 4 – S = 350 m²/kg

modern theoretical concepts [72], the process of structure formation of binder systems can be divided into three periods:

1. induction, immediately after mixing the binder with water;
2. period of coagulation structure formation (setting);
3. period of crystallization structure formation.

The plastic strength of the pastes was determined using a conical plastometer. The depth of the cone immersion was recorded with an indicator with a division value of 0.1 mm. The plastic strength was calculated using the following formula:

$$P_p = 0{,}096\ P/h^2, \qquad (3.12)$$

where, P – weight of the cone with the load, H;

h – cone immersion depth in mm.

Comparison of the duration of the first section of the plastograms (Figs. 3.10...3.11) with the duration of the setting time of the pastes (Tables 3.8, 3.9) shows their correspondence.

The inflection points on the plastograms, separating the first and second sections, approximately coincide with the end of the setting of the pastes. Factors that contribute to the acceleration of the setting

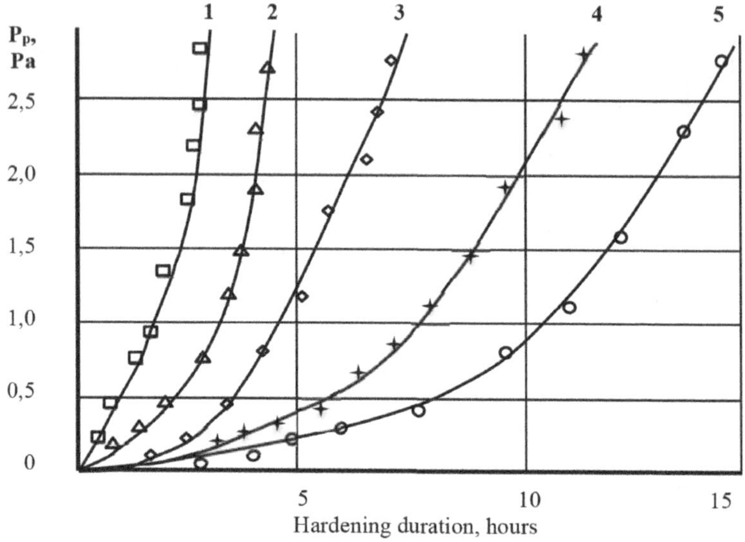

Fig. 3.10 Plastic strength of pastes based on three-component binders of the composition Portland cement-slag-dust (D1) 2:1:1 (by mass): 1 – S = 630 m²/kg, 1% SP-1; 2 – S = 590 m²/kg; 3 – S = 630 m²/kg, 1% SP-1, gypsum stone – 3%, 4 – S = 360 m²/kg, 1% SP-1; 5 – S = 330 m²/kg

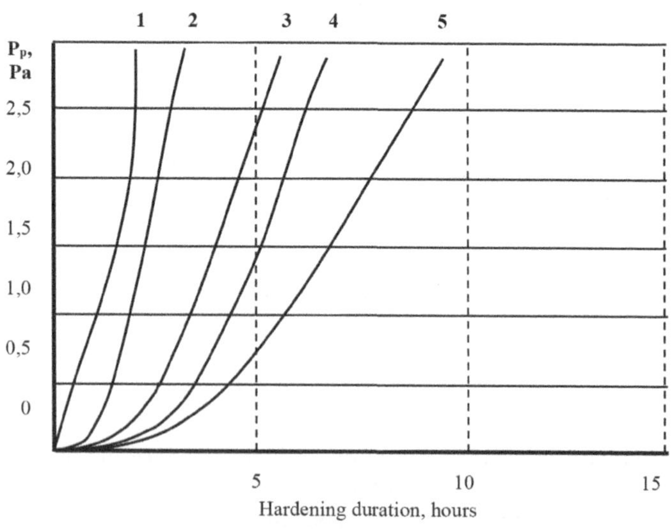

Fig. 3.11 Plastic strength of pastes based on three-component binders of the composition Portland cement-slag-dust (D2) 2:1:1 (by mass): 1 – S = 630 m²/kg, 1% SP-1; 2 – S = 620 m²/kg; 3 – S = 630 m²/kg, gypsum stone – 3%; 4 – S = 350 m²/kg, SP-1% – 1%; 5 – S = 320 m²/kg

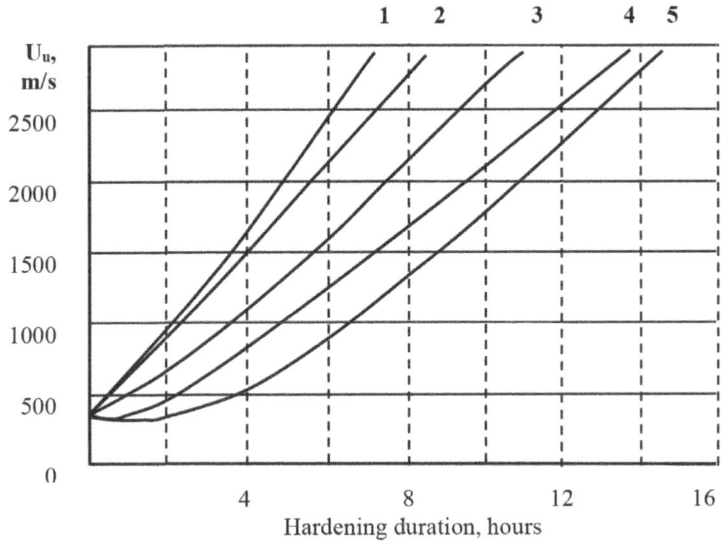

Fig. 3.12 Ultrasonic wave propagation velocity in hardening pastes based on composite binders cement:slag:dust (D2) 2:1:1 (by mass): 1 – S = 630 m^2/kg, 1% SP-1; 2 – S = 590 m^2/kg; 3 – S = 360 m^2/kg, 1% SP-1; 4 – S = 330 m^2/kg; 5 – S = 630 m^2/kg, 1% SP-1, gypsum stone – 3%

time, the decrease of the water-binding ratio with the introduction of a superplasticizer, and the increase in the specific surface area also contribute to the reduction of the period of coagulation structure formation. These same factors contribute to the strengthening of the coagulation structure and the intensification of the process of formation of the crystallization structure. Pastes based on three-component binders in the Portland cement-slag-dust system, obtained by grinding to a specific surface area of over 500 m^2/kg, are characterized by a very short section on plastograms, which corresponds to coagulation structure formation. The additional introduction of dihydrate gypsum into pastes obtained on the basis of binders with a high specific surface area contributes to the extension of the period corresponding to the stability of the coagulation structure. However, after the end of setting, such pastes are characterized by a more intensive formation of the crystallization framework of the hardening structure.

Fluctuations in the alkali content of the dust, as noted earlier, lead to a certain acceleration of the initial of setting time, but the final of setting time and, accordingly, the period of existence of the coagulation structure of the pastes does not change noticeably.

The initial period of structure formation of pastes based on composite binders was also studied by the pulsed ultrasonic method [73]. The measurement of the transit time of ultrasonic waves was carried out using the device at frequencies of 60 kHz with an accuracy of 0.25–0.5 μs. The measurement was carried out on freshly prepared cube samples with dimensions of 7.07×7.07×7.07 cm in Plexiglas molds during the hardening of the samples under normal conditions. The time τ_u and the speed of propagation of ultrasonic waves υ_u were determined by the formulas:

$$\tau_u = f_2 - f_1 \tag{3.13}$$

$$\upsilon_u = l_b \, \tau_u \, (f_2 - f_1) \tag{3.14}$$

where, l_b – basic sample size (l_b=7,07 *cm*);
f_2–f_1 – frequency difference.

On the curves of the speed of propagation of ultrasonic waves, the initial period of formation of the coagulation structure is characterized by a horizontal section in Fig. 3.12, the length of which is close in magnitude to the duration of the onset of solidification.

As in the curves of changes in plastic strength over time, the rate of formation of the crystallization structure can be related to the angle of inclination of the curves of the speed of passage of ultrasonic waves through the hardening pastes. 14...18 hours after mixing, the speed of ultrasonic waves increases by 4–5 times. The value of v_u, as well as P_p, is most significantly influenced by the specific surface area of composite binders and the water-binding ratio.

3.4 The Influence of Composition and Specific Surface Area of Composite Binders on Their Strength

As shown earlier, adding clinker kiln dust to Portland cement is advisable in a composition with blast furnace granulated slag, which not only eliminates the negative features of dust as an additive to cement but is also activated under its influence.

To study the influence of the content and ratio of the main components, as well as the fineness of grinding, on the strength indicators of cement-dust-slag binders (CDSB) experiments algorithmized according to the typical B_4 plan [22] were implemented and mathematical models of the compressive and flexural strength of standard mortars at the age of 2, 7, and 28 days were obtained.

The initial components in the implementation of the experimental plan B_4 were Portland cement of the I type (CEM I) and dust removal

(Tables 3.5, 3.6) of clinker kilns (D1), as well as blast furnace granulated slag of the Kryvyi Rih Metallurgical Plant (Ukraine). As superplasticizer additives SP-1 and Sika VC225 were used. Grinding of binders was carried out in a laboratory mill. During grinding, the superplasticizer additive was introduced in powder form.

The experimental design conditions are given in Table 3.12, the obtained experimental results and experimental statistical models are given in Tables 3.13, 3.14.

Table 3.12 Conditions for planning experiments when studying the influence of composition and specific surface area on the strength of composite binders.

Factors		*Levels of Variation*			*Variation Interval*
Natural View	*Coded View*	*– 1*	*0*	*+1*	
$(D_1 + S_1)$, %	X_1	10	35	60	25
$\frac{D_1}{D_1 + S_1}$	X_2	0	0,5	1	0,5
S, m^2/kg	X_3	300	450	600	150
SP, %	X_4	0	1	2	1

Note: D_1 – rotary kiln dust, Sl – granulated blast furnace slag,
S – specific surface area, SP – superplasticizer SP-1.

Table 3.13 Experimental results obtained when studying the influence of the composition of CDSB on their strength properties.

Point No.	X_1	X_2	X_3	X_4	*Compressive Strength, MPa*			*Bending Strength, MPa*		
					2 Days	*7 Days*	*28 Days*	*2 Days*	*7 Days*	*28 Days*
1	1	1	1	1	9,4	21,7	30,1	1,88	3,62	3,77
2	1	1	1	– 1	3,9	14,6	21,6	0,79	2,43	2,7
3	1	1	– 1	1	6,7	14,0	20,1	1,34	2,33	2,51
4	1	1	– 1	– 1	2,8	9,3	14,8	0,55	1,55	1,85
5	1	– 1	1	1	15,4	40,7	58,6	3,08	6,79	7,32
6	1	– 1	1	– 1	10,9	33,5	47,8	2,19	5,58	5,97
7	1	– 1	– 1	1	12,2	29,9	44,3	2,43	4,99	5,53
8	1	– 1	– 1	– 1	9,2	25,1	36,7	1,85	4,19	4,59
9	– 1	1	1	1	22,5	47,0	66,9	4,49	7,84	8,37

Point No.	X_1	X_2	X_3	X_4	*Compressive Strength, MPa*			*Bending Strength, MPa*		
					2 Days	*7 Days*	*28 Days*	*2 Days*	*7 Days*	*28 Days*
10	– 1	1	1	– 1	17,5	39,9	58,4	3,51	6,65	7,3
11	– 1	1	– 1	1	16,8	37,3	53,1	3,35	6,22	6,64
12	– 1	1	– 1	– 1	13,4	32,6	47,8	2,67	5,44	5,98
13	– 1	– 1	1	1	24,9	50,2	72,1	4,99	8,36	9,02
14	– 1	– 1	1	– 1	21,0	42,9	61,3	4,2	7,16	7,67
15	– 1	– 1	– 1	1	18,7	37,4	54,1	3,74	6,23	6,76
16	– 1	– 1	– 1	– 1	16,3	32,6	46,5	3,26	5,43	5,81
17	1	0	0	0	16,9	35,2	49,8	3,39	5,87	6,22
18	– 1	0	0	0	27,0	51,6	73,1	5,40	8,6	9,13
19	0	1	0	0	15,7	25,6	36,1	3,15	4,27	4,52
20	0	– 1	0	0	20,2	35,1	49,7	4,04	5,85	6,21
21	0	0	1	0	19,3	39,1	55,1	3,86	6,51	6,89
22	0	0	– 1	0	15,6	30,0	42,7	3,12	5,01	5,34
23	0	0	0	1	21,7	38,5	54,5	4,33	6,42	6,81
24	0	0	0	– 1	17,8	32,6	46,4	3,55	5,43	5,8

Analysis of polynomial models (Table 3.14) obtained as a result of processing experimental data allows us to trace the clearly expressed nonlinear nature of the influence of the studied factors on the strength of composite binders and to find their optimal values (Figs. 3.4, 3.5).

Table 3.14 Mathematical models of strength indicators CDSB.

Strength Parameters		*Mathematical Models*	
Compressive strength, MPa	2 days	$y_1 = 21{,}09 - 5{,}04X_1 - 2{,}24X_2 + 1{,}85X_3 + 1{,}96X_4 + 0{,}88X_1^2 - 3{,}12X_2^2 - 3{,}62X_3^2 - 1{,}38X_4^2 - 0{,}88X_1X_2 - 0{,}75X_1X_3 + - 0{,}13X_2X_3 - 0{,}25X_2X_4 + 0{,}38X_3X_4$	(3.15)
	7 days	$y_2 = 37{,}35 - 8{,}19X_1 - 4{,}74X_2 + 4{,}52X_3 + 2{,}98X_4 + 6{,}04X_1^2 - 7X_2^2 - 2{,}8X_3^2 - 1{,}8X_4^2 - 3{,}97X_1X_2 - 0{,}5X_1X_3 - 0{,}78X_2X_3 - 0{,}03X_2X_4 + 0{,}61X_3X_4$	(3.16)
	28 days	$y_3 = 52{,}6 - 11{,}65X_1 - 6{,}78X_2 + 6{,}22X_3 + 4{,}03X_4 + 8{,}82X_1^2 - 9{,}68X_2^2 - 3{,}68X_3^2 - 2{,}18X_4^2 - 5{,}81X_1X_2 - 0{,}94X_1X_3 - 1{,}06X_2X_3 - 0{,}56X_2X_4 + 0{,}81X_3X_4$	(3.17)

Strength Parameters		*Mathematical Models*	
Bending strength, MPa	2 days	$y_4 = 4{,}22 - 1{,}02X_1 - 0{,}45X_2 + 0{,}37X_3 + 0{,}39X_4 + 0{,}17X_1^2 - 0{,}63X_2^2 - 0{,}73X_3^2 - 0{,}28X_4^2 - 0{,}18X_1X_2 - 0{,}15X_1X_3 + 0{,}03X_1X_4 - 0{,}03X_2X_3 - 0{,}05X_2X_4 + 0{,}08X_3X_4$	(3.18)
	7 days	$y_5 = 6{,}23 - 1{,}38X_1 - 0{,}79X_2 + 0{,}76X_3 + 0{,}5X_4 + 0{,}99X_1^2 - 1{,}12X_2^2 - 0{,}45X_3^2 - 0{,}31X_4^2 - 0{,}66X_1X_2 - 0{,}08X_1X_3 + 0{,}01X_1X_4 - 0{,}13X_2X_3 - 0{,}005X_2X_4 + 0{,}1X_3X_4$	(3.19)
	28 days	$y_6 = 6{,}58 - 1{,}47X_1 - 0{,}85X_2 + 0{,}78X_3 + 0{,}51X_4 + 1{,}09X_1^2 - 1{,}22X_2^2 - 0{,}47X_3^2 - 0{,}29X_4^2 - 0{,}73X_1X_2 - 0{,}12X_1X_3 + 0{,}01X_1X_4 - 0{,}13X_2X_3 - 0{,}07X_2X_4 + 0{,}1X_3X_4$	(3.20)

The influence of such factors as the part of dust in the mass of the binder and its specific surface area is extreme. For the specific surface area (Figs. 3.6, 3.7), it is especially felt at 2-day age in the range of 450–500 m^2/kg.

Increasing the content of dust-slag filler affects the strength of binders differently depending on the duration of hardening and the ratio of dust to slag (Figs. 3.13, 3.14). At an early age, increasing the dust content in the binder composition almost linearly reduces its strength. The strength decreases less, although significantly, when blast furnace slag and a combination of slag and dust are added. At the same time, when combining slag and dust in a 1:1 ratio by weight, the strength decrease occurs approximately the same as when only slag is added. At 28 days of age, increasing the dust content in the binder from 10 to 30% reduces the compressive strength by 35%, and at 50%, almost twice. The sharp drop in the strength of cement filled with dust is largely explained by the increase in its water consumption (Table 3.15). Increasing the content of dust-slag filler in the dust:slag ratio of 1:1 by weight as well as slag alone from 10 to 50% causes a decrease in strength by approximately only 30% (Figs. 3.13, 3.14). Increasing the proportion of dust in the dust-slag filler from 0 to 50% practically does not cause an additional decrease in strength.

Reducing the water consumption of binders by adding a superplasticizer and increasing the specific surface area allows for a significant increase in their strength. Additional grinding cement with a dust content of 50–60% and the addition of SP-1 to a specific surface

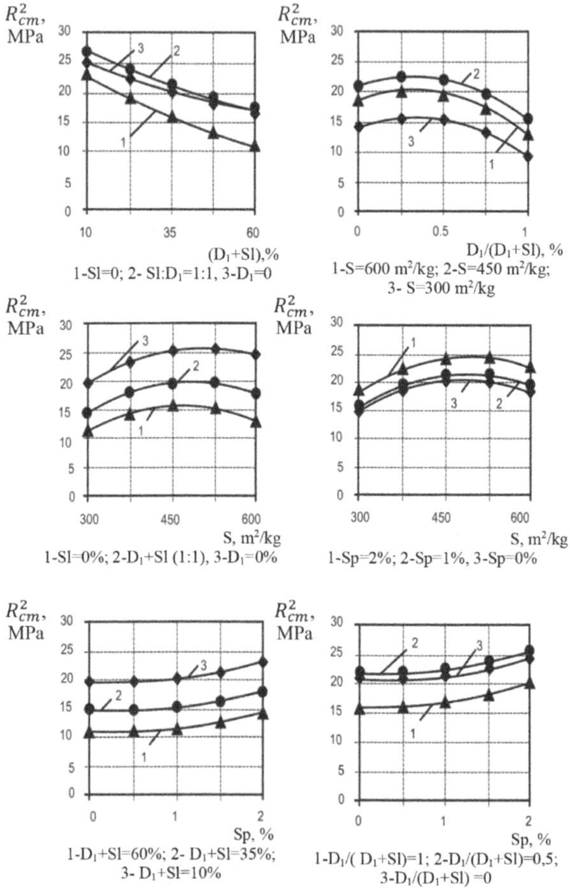

Fig. 3.13 Dependence of compressive strength of composite cement-slag binder at the age of 2 days.

Note: The factors listed in Table 3.12, the influence of which is not considered in Fig. 3.13, are at the basic (zero) level.

area of 450–500 m^2/kg allows to increase the compressive strength from 30 to 40 MPa. With a content of a complex additive of cement dust-blast slag of 50–60% in a ratio of 1:1 by weight, under the same conditions, it is possible to obtain a binder with a compressive strength at the age of 28 days of 50–60 MPa (Table 3.13, Figs. 3.13, 3.14).

Replacing the naphthalene-formaldehyde superplasticizer with a polycarboxylate one allows, with a content of 50% complex slag-dust active filler, to increase the grade strength of the composite binder to 70 MPa (Table 3.16).

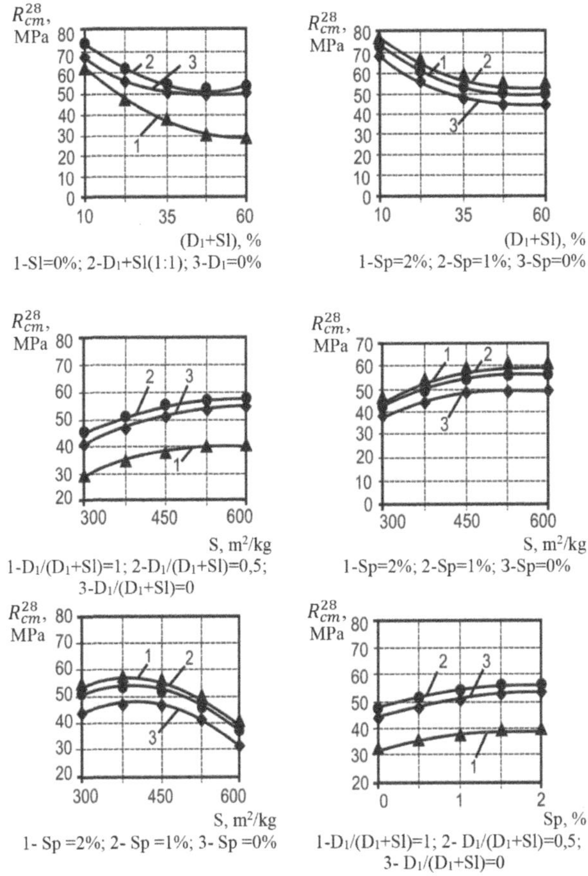

Fig. 3.14 Dependence of compressive strength of composite binder at the age of 28 days. *Note:* The factors listed in Table 3.12, the influence of which is not considered in Fig. 3.14, are at the basic (zero) level.

The resulting composite binders with low water consumption are fast-setting, their 2-day strength is not less than 50% of the 28-day strength, and the ratio of compressive strength to flexural strength for optimal compositions is within the usual limits typical of Portland cement.

Taking into account the possibility of fluctuations in the content of alkaline compounds and free lime in the dust, their influence on the strength indicators of low water consumption CDSB containing dust and granulated blast furnace slag in an optimal ratio was determined. For this purpose, additional experiments were conducted using dust D_2, which contained an increased amount of alkaline oxides (5.6%) and free lime

Table 3.15 Water consumption of composite binders and mortars.

No	*Binder Composition, %*				*S, m^2/kg*	*Normal Consistency, %*	*Water-binding Ratio of the Mortars (cone spread 115 mm)*
	Cement	*Slag*	*Dust*	*Super-plasticizer SP-1*			
1	65	-	35	-	310	32,3	0,41
2	40	-	60	-	310	36,5	0,43
3	65	-	35	1	320	28,3	0,40
4	65	-	35	2	320	27,8	0,40
5	65	35	-	-	310	27,1	0,40
6	40	60	-	-	310	28,2	0,40
7	40	60	-	1	325	22,2	0,37
8	40	60	-	2	325	21,4	0,36
9	50	25	25	-	320	29,1	0,41
10	50	25	25	1	330	22,7	0,36
11	50	25	25	2	330	21,6	0,35
12	50	25	25	1	460	21,7	0,35
13	50	25	25	1	580	22,1	0,36
14	50	25	25	0,6*	450	19,5	0,33
15	50	25	25	0,6*	590	19,8	0,33

Notes: 1. When grinding the binder of compositions Nos. 14 and 15 the polycarboxylate type superplasticizer Sica VC 225 was added. 2. Gypsum and superplasticizer introduced into the cement are accounted for by more than 100%.

Table 3.16 Strength of composite binders with the addition of superplasticizer Sika VC 225.

No	*Binder Composition, %*				*S, m^2/kg*	*Strength, MPa*					
						Bend, Days			*Compression, Days*		
	Cement	*Slag*	*Dust*	*SP Sika*		*2*	*7*	*28*	*2*	*7*	*28*
1	50	25	25	0,3	320	3,8	4,7	5,6	28	36	42
2	50	25	25	0,3	450	4,7	5,8	6,5	31	44	58
3	50	25	25	0,3	580	5,1	6,3	7,1	39	51	64
4	50	25	25	0,6	320	4,8	5,7	6,4	35	43	56
5	50	25	25	0,6	450	5,9	6,8	7,5	49	61	72
6	50	25	25	0,6	580	6,2	7,1	7,9	51	63	75
7	50	25	25	0,7	450	6,0	6,9	7,7	50	62	73

Note: Gypsum and superplasticizer added to cement are counted as more than 100%.

(5.9%). The binder consisted of 60% dust and slag in a ratio of 1:1. The results of the experiments are given in Table 3.17.

As follows from the data given in Table 3.17, a slight increase in the content of R_2O and CaO_{free} in the dust causes a slight tendency to increase the strength indicators of composite binders with low water consumption, which is associated with additional activation of the slag component.

Table 3.17 The influence of the content of alkalis R2O and free lime CaOfree in the dust on the strength of the CDSB.

Dust Sample	*Content SP*	*Specific Surface Area m²/kg*	*Strength, MPa*					
			Bend, Day			*Compression, Day*		
			2	*7*	*28*	*2*	*7*	*28*
D_1-1 (R_2O = 2,9%, CaO_{free} = 3,1%)	SP-1–1%	450	3,4	5,9	6,2	16,9	35,2	49,8
	SP-1–2%	450	3,5	6,1	6,5	17,6	36,4	51,6
	Sika–0,6%	450	5,9	6,8	7,5	49,0	61,0	72,0
D_1-2 (R_2O = 5,6%, CaO_{free} = 5,9%)	SP-1–1%	450	3,6	6,5	6,9	18,6	39,4	53,3
	SP-1–2%	450	3,8	6,7	7,1	19,6	39,7	55,7
	Sika –0,6%	450	6,3	7,2	7,8	51,0	63,0	74,0

3.5 The Influence of the Composition of Dry Mixes Containing Clinker Kiln Dust on the Properties of Building Mortars

Dry construction mixtures were prepared by thoroughly mixing Portland cement, fly ash, ground granulated blast furnace slag, and sand in a laboratory mixer with the addition of a powdered superplasticizer additive, SP-1. The slag was ground to a specific surface area S = 310–330 m^2/kg in a laboratory ball mill. The blown dust corresponded to the chemical composition of the D_1 dust given in Table 3.5. To achieve the required homogeneity of the mixtures, they were mixed in two stages: at the first stage, with the binder components, at the second, with the introduction of sand.

To determine the influence of the composition of dry mixtures on the normalized properties of mortars, experiments were performed, algorithmized according to a typical semi-replica plan of a full-factorial experiment with four factors varying [22]. The experimental design conditions are given in Table 3.18.

Table 3.18 Conditions for planning experiments on the influence of the composition of dry mixtures on the properties of mortars.

Factors		*Levels of Variation*			*Variation Interval*
Natural View	*Coded View*	*– 1*	*0*	*+1*	
Cement consumption (C), kg/t	X_1	100	150	200	50
Slag consumption (Sl), kg/t	X_2	50	75	100	25
Dust consumption (D), kg/t	X_3	50	75	100	25
Consumption of superplasticizer (SP-1), % of the binder mass	X_4	0	0,5	1	0,5

The consumption of quartz sand at each point of the matrix was determined from the condition: Sand = 1000-C-Sl-D_1. Sand was used with a maximum particle size of 1.25 mm (M_f = 1.3). Construction mixtures at each point of the matrix were obtained by mixing the dry mixture while bringing it to a mobility of 8...10 cm immersion of the standard cone.

The initial parameters were determined as the compressive strength of mortars for samples hardening in air-dry conditions, adhesion strength to the base, and frost resistance. The compositions of the studied mixtures and the results of experiments on determining the properties of mortars based on them are given in Table 3.19.

Additionally, the yield coefficient of the mortar mixture (l/kg) and the binder-water ratio necessary to achieve the specified workability of the mortar mixture were also found. As a result of statistical processing of experimental data, incomplete quadratic regression equations of the studied parameters on the varying factors were obtained, given in the coded variables in Table 3.20. Along with the linear effects of the influence of individual factors, equations (3.21...3.26) also take into account the effects of their pairwise interaction.

Analysis of equations describing the influence of dry mixture composition factors on the compressive strength of mortars shows (Figs. 3.15, 3.16) that in the studied range of composition factors changes, it is possible to obtain mortars with a 28-day compressive strength of 10…30 MPa. By 7 days of age, the strength of the mortars reaches 70% of the standard 28-day strength. Strength naturally decreases when the composition of the mixtures changes, which leads to a decrease in the binder-water ratio. In this regard, the impact of cement consumption and superplasticizer addition is particularly significant.

In the regression equations, for both 7-day and 28-day strength, the most significant linear effects are found for factors X_1 and X_4, i.e.,

Table 3.19 Compositions of dry mixes and properties of mortars based on them.

No	*Composition of Dry Mix, kg/t*				*Mortars Yield Coefficient (Kyield), l/kg*	*Binder-water Ratio (Binder/W)*	*Compressive Strength, MPa*		*Bond Strength to the Base, MPa (Rbond)*	*Frost Resistance, Cycles (F)*
	Cement (C)	*Slag (Sl)*	*Dust (D1)*	*Sand*			*7 Day* (R^7_{cm})	*28 Day* (R^{28}_{cm})		
1*	200	100	100	600	0,7	2,94	21,8	27,6	0,82	100
2	200	100	50	650	0,5	2,33	17,4	23,5	0,61	75
3	200	50	100	650	0,5	2,27	8,5	14,2	0,45	50
4*	200	50	50	700	0,65	2,63	21,0	30,0	0,74	75
5	100	100	100	700	0,6	1,92	7,1	13,5	0,43	35
6*	100	100	50	750	0,65	2,08	6,7	13,8	0,34	50
7*	100	50	100	750	0,65	2,5	8,1	15,4	0,51	50
8	100	50	50	800	0,55	1,35	3,4	7,4	0,25	25

*In the mixture of compositions 1, 4, 6, 7, the superplasticizer SP-1 was introduced in the amount of 4; 3; 2.5; 2.5 kg/t, respectively.

Table 3.20 Mathematical models of the properties of mortars based on dry mixtures.

Properties of the Solution	*Output Parameter*	*Regression Equation*	
Compressive strength at the age of 7 days (MPa)	R^{7}_{cm}	$R^{7}_{cm} = 11{,}8 + 5{,}45X_1 + 1{,}5X_2 - 0{,}38X_3 + 2{,}65X_4 + 0{,}93X_1X_2 - 1{,}65X_1X_3 + 1{,}558X_1X_4 + 1{,}58X_2X_3 - 1{,}65X_2X_4 + 0{,}92X_3X_4$	(3.21)
Compressive strength at 28 days (MPa)	R^{8}_{cm}	$R^{28}_{cm} = 18{,}0 + 5{,}81X_1 + 1{,}59X_2 - 0{,}33X_3 + 3{,}69X_4 + 0{,}14X_1X_2 - 2{,}6X_1X_3 + 1{,}29X_1X_4 + 1{,}29X_2X_3 - 2{,}6X_2X_4 + 0{,}14X_3X_4$	(3.22)
Adhes ive strength at 28 days (MPa)	R_{ad}	$R_{ad} = 0{,}52 + 0{,}14X_1 + 0{,}03X_2 - 0{,}03X_3 + 0{,}08X_4 + 0{,}03X_1X_2 - 0{,}05X_1X_3 + 0{,}04X_1X_4 + 0{,}04X_2X_3 - 0{,}5X_2X_4 + 0{,}14X_3X_4$	(3.23)
Frost resistance, cycles	F	$F = 57{,}5 + 17{,}5X_1 + 7{,}5X_2 + 1{,}25X_3 + 11{,}25X_4 + 5X_1X_2 - 1{,}25X_1X_3 + 1{,}25X_1X_4 + 1{,}3X_2X_3 - 1{,}3X_2X_4 + 5X_3X_4$	(3.24)
Binder-water ratio	Bind/W	$\text{Bind}/W = 2{,}25 + 0{,}29X_1 + 0{,}06X_2 + 0{,}15X_3 + 0{,}28X_4 + 0{,}028X_1X_2 - 0{,}09X_1X_3 - 0{,}04X_1X_4 + 0{,}04X_2X_3 - 0{,}09X_2X_4 + 0{,}028X_3X_4$	(3.25)
Mortar mix yield coefficient	K_y	$K_y = 0{,}606 + 0{,}06X_1 + 0{,}031X_2 + 0{,}006X_3 + 0{,}031X_4 + 0{,}006X_1X_2 + 0{,}006X_1X_3 - 0{,}007X_1X_4 + 0{,}005X_2X_3 + 0{,}007X_2X_4 + 0{,}006X_3X_4$	(3.26)

cement and superplasticizer consumption. An increase in the slag content in the mixture can cause both a slight increase and a decrease in the 28-day strength of the mortars. The increase in the strength of mortars with increasing slag content is characteristic for a reduced cement content (Figs. 3.15, 3.16) and, conversely, in mixtures containing an increased cement consumption, a decrease in strength is observed with increasing slag content.

Reducing the dust content within selected limits can also reduce or have no effect on strength. Some increase in the strength of mortars with

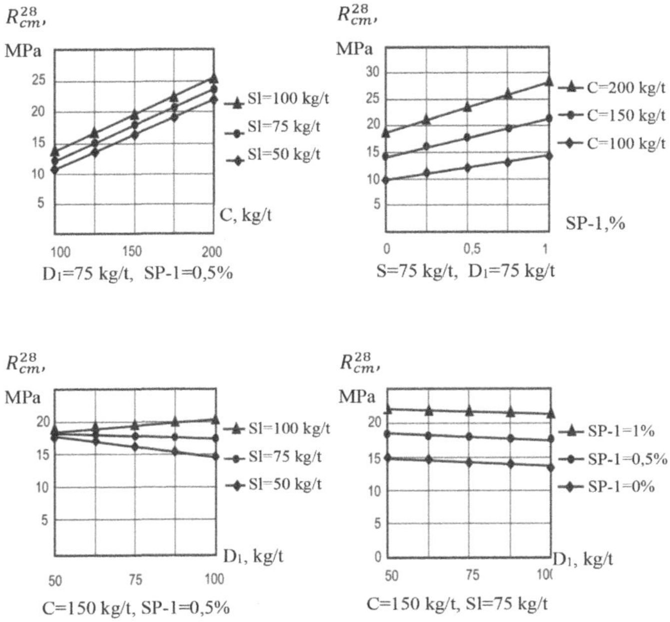

Fig. 3.15 The influence of the composition of the dry mixes on the compressive strength of mortars at the age of 28 days.

increasing dust content is observed with a simultaneous increase in slag content. With insufficient slag content in mortar mixtures, an increase in dust content leads to a decrease in strength. A decrease in strength is also possible in the absence of a superplasticizer additive in the mortar mixture (Figs. 3.15, 3.16).

In relatively low-cement mortars, the joint introduction of slag and dust as active mineral additives in amounts of up to 200 kg/m^3 each leads to a significant increase in both 7- and 28-day compressive strength.

The influence of the composition of the dry mixes on the binder-water ratio and the mortars' output coefficient is shown in Figs. 3.16, 3.17.

Experimental-statistical models can be used to design the compositions of both dry mixtures and mortars based on them. The composition of mixtures that provide a given strength is easily selected using equations (3.21, 3.22) and taking into account possible resource limitations. For this purpose, it is possible to use a nomogram (Fig. 3.18).

To ensure a set of required properties, for example, 28-day compressive strength, adhesion strength to the base, and frost resistance, a system of three corresponding equations is solved (the consumption

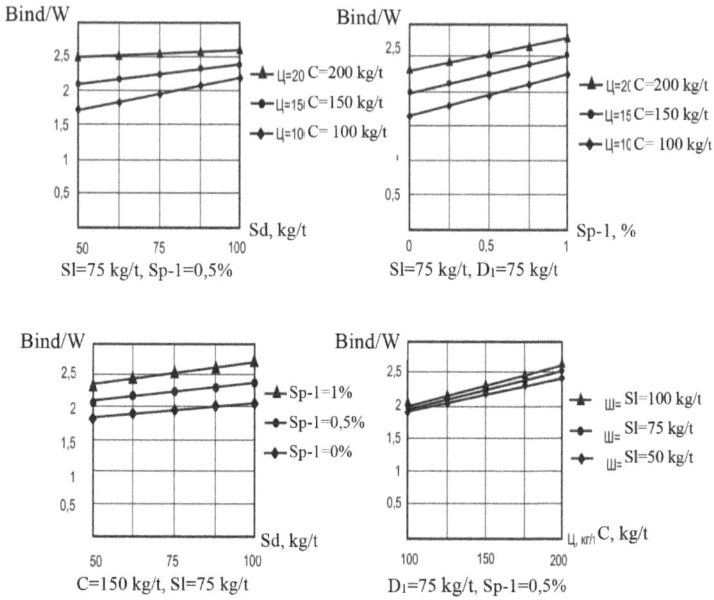

Fig. 3.16 Effect of dry mix composition on the binder-water ratio.

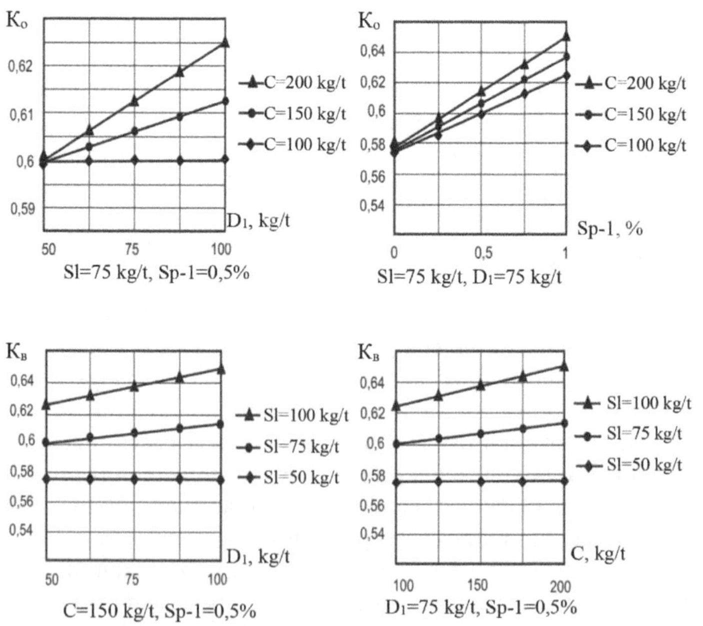

Fig. 3.17 Influence of the composition of the dry mixture on the mortars yield coefficient.

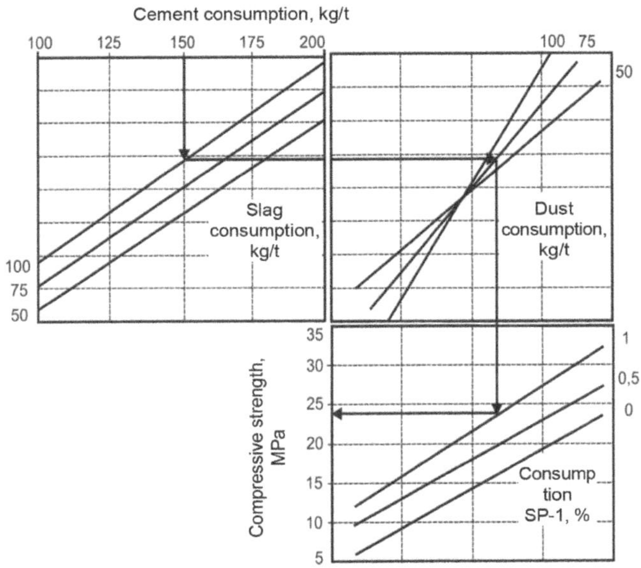

Fig. 3.18 Nomogram for calculating the composition of dry building mix for obtaining masonry mortar for 3D printing.

of superplasticizer X_4 is given) and the required consumption (kg/t) of cement, slag, and dust (X_1, X_2, and X_3) is found.

When moving from the composition of the dry mixture (kg/t) to the composition of the mortars by calculation using the corresponding dependence (Eq. 3.27), we find the value of the yield coefficient and recalculate the consumption of each component:

$$P_{P.C.} = \frac{P_{C.C}}{K_o}, \tag{3.27}$$

where, $P_{P.C.}$ – consumption of mortar components, kg/m^3;

$P_{C.C.}$ – consumption of dry mix component, kg/t;

K_O – output coefficient, m^3/t.

If necessary, it is possible to analyze the calculated compositions that ensure the minimum price of the mixture:

$$Pr_m = Pr_C C + Pr_{sl} Sl + Pr_d D + Pr_{sd} Sd + Pr_{sp} SP, \tag{3.28}$$

where, Pr_m – total cost of materials per 1 ton of dry mix or 1 m^3 of mortar mix;

Pr_c, Pr_{sl}, Pr_d, Pr_{sd}, Pr_{sp} – the cost of cement, blast furnace slag, dust, sand, and superplasticizer;

C, Sl, D, Sd, Sp – specific costs of the relevant components.

Table 3.21 shows the compositions of dry mixes of different grades by strength, calculated using equations that ensure the minimum possible cement consumption, taking into account the range of factors variation adopted in Table 3.18.

Table 3.21 Estimated compositions of dry mixes and solutions.

Composition No.	*Mortar Brand by Strength in 1 MPa*	*The Composition of the Dry Mixture*					*Output Coefficient*	*Binder/water Ratio*
		Cement, kg/t	*Slag, kg/t*	*Dust, kg/t*	*Sand, kg/t*	*Superplasticizer SP-1,%*		
1	50	100	75	50	775	0	0,53	1,43
2	75	100	75	50	775	0,3	0,54	1,60
3	100	200	75	50	675	0,5	0,65	2,48
4	150	200	50	100	650	1	0,67	2,91
5	200	105	100	100	695	1	0,62	2,47
6	250	140	100	50	710	1	0,64	2,29

Figure 3.19 shows the calculated dependences characterizing the relationship between the strength of mortars and the binder-water ratio.

The dependencies strength of mortars can be interpreted by formulas like:

$$R_m = K(\text{Bind}/W - b), \tag{3.29}$$

where, K and b – coefficients that take into account the characteristics of the materials used. For the materials used at the hardening age of the mortars 7 days $K = 12.7$, $b = 1.32$; 28 days – $K = 15$, $b = 1.05$.

The coefficient K should also reflect to some extent the influence of the activity of the binders used. When taking into account the influence of the activity of the binders, formula (Eq. 3.29) will have the form:

$$R_m = AR_{\text{binder}}(\text{Bind}/W - b), \tag{3.30}$$

The calculated and experimental values of the strength of mortars, calculated by formula (3.29), are given in Table 3.22.

An important indicator of the quality of mortars for 3D printing is the measured pull-off strength, which is determined by their cohesive and adhesive properties. The main cohesive parameter of the mortar is its compressive strength. The adhesion of cement mortars to the bases is in a complex dependence on many factors, including the dispersion of the binder, the wettability of the base by the mortar mixture, etc.

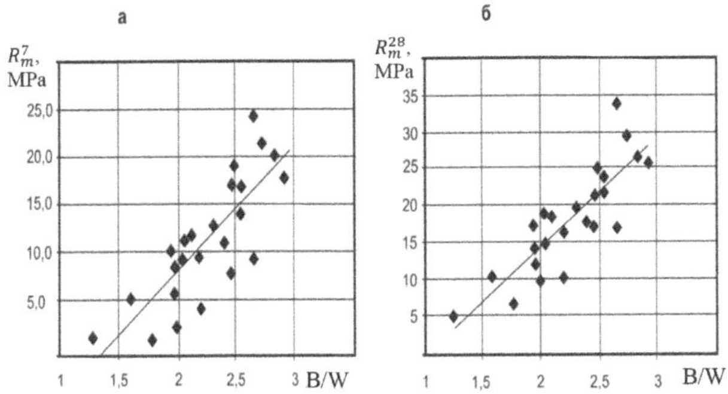

Fig. 3.19 Relationship between the compressive strength of mortars and the binder-water ratio at the age: *a* – 7 days; *b* – 28 days.

Table 3.22 Calculated and experimental values of mortar strength.

Composition Number According to Table 3.21	*Calculated Values of 28-day Compressive Strength, MPa*	*Experimental Values of 28-day Compressive Strength, MPa*	*Relative Error, %*
1	5,7	6,3	10
2	8,3	7,7	15
3	21,5	24,1	12
4	27,9	25,4	9
5	21,3	23,6	11
6	18,6	16,7	11

The main role in the formation of adhesive strength is given to the ratio of the values of the surface energy of the adhesive (W_{ad}) and the substrate (W_{sub}). In this case, the condition $W_{ad} < W_{sub}$ is mandatory. The value of the surface energy is determined by the features of their structure, chemical composition, dispersion, etc. [74].

Wetting of the base mortars and increasing adhesion strength is facilitated by reducing surface energy when creating an adsorption-active environment due to surfactant additives. The influence of the adsorption-active medium increases with increasing filler dispersion and its concentration, which is associated with an increase in the interfacial surface and, accordingly, surface energy.

The adhesive strength of mortars on composite binders, including clinker dust and granulated blast furnace slag for all compositions in the studied area of change of factors exceeds the regulatory requirements. It increases significantly with an increase in the volume of the binder component in dry mixtures (Fig. 3.20). A decrease in the water-

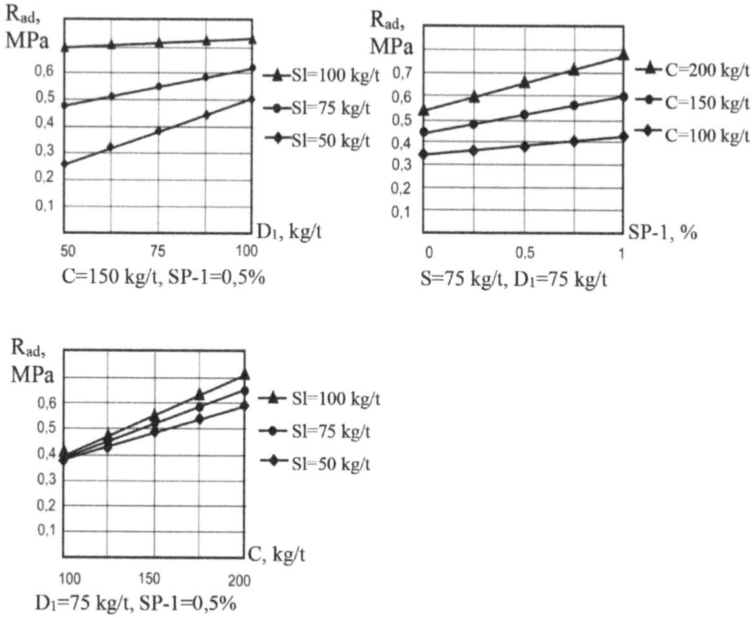

Fig. 3.20 The influence of the composition of the dry mixture on the adhesive strength of mortars.

binding ratio also has a positive effect. In addition, the inclusion n of a superplasticizer, which belongs to the hydrophilic surfactants, should contribute to improving the wetting of the substrate surface and increasing the adhesion strength.

The addition of fly ash to the mortar mixture has a positive effect on the adhesive strength. The degree of its influence increases with increasing specific content in the mortar mixture, while simultaneously increasing the slag content and decreasing the cement content (Fig. 3.20). This result can be explained by the improvement of the binding properties of the dust-slag composition when the optimal ratio of components is achieved. At the same time, the influence of the cement dust-slag binder on the adhesion of the mortars to the base becomes especially significant when there is a lack of cement paste (Fig. 3.20).

The positive effect of dust as a highly dispersed component of the dry mix, which increases its water demand, on the adhesive strength increases with the introduction of a superplasticizer additive, which reduces the negative effect of dust on the water content. Enhancing the dust positive effect on the adhesive strength of the mortars to the base is reflected by

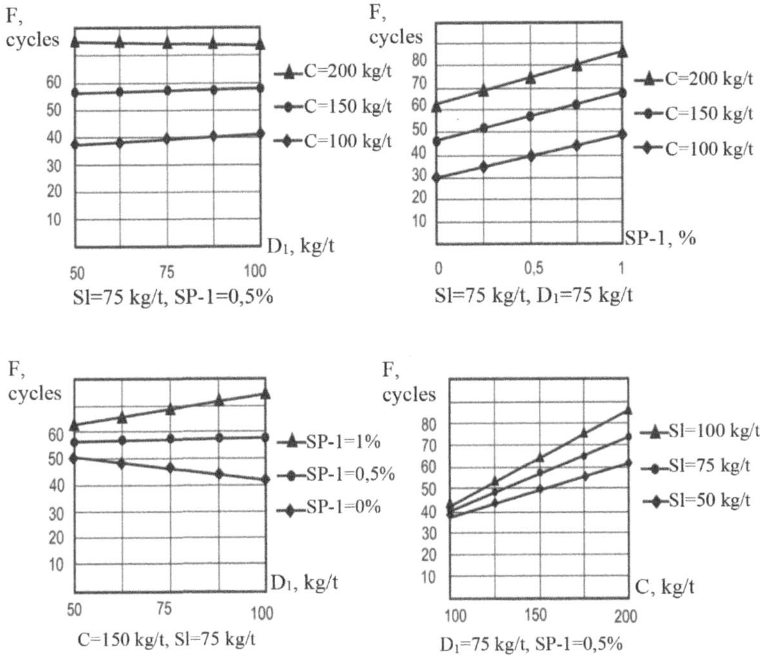

Fig. 3.21 Influence of the composition of the dry mix on the frost resistance of solutions.

the presence in the corresponding regression equation (Table 3.20) of the most significant effect of the interaction of factors X_3 and X_4.

The frost resistance of mortars based on composite binders, as follows from the analysis of regression equations (Table 3.20), is in the range of values allowed for masonry mortars and is mainly determined by factors affecting the binder-water ratio, i.e., cement consumption and the content of the superplasticizer additive (Fig. 3.21). The effect of the dust additive on the frost resistance of concrete depends on how much its introduction changes the water content and, accordingly, the water-binding ratio. If the mixture does not contain a superplasticizer additive, the introduction of an increased content of dust leads to a decrease in frost resistance, in the presence of the SP-1 additive, frost resistance may practically not change or even increase slightly.

3.6 Technological Properties of Mortar Mixtures

The technological suitability of mortar mixtures for 3D printing is largely determined by their workability.

In practice, the task arises of specifying the required amount of water depending on the composition of the mixture and the required workability. Knowing the required binder-water ratio to achieve the required strength, it is possible to adjust or select the composition of the mixture accordingly so that it provides both the specified workability and the strength of the mortars.

To determine the effect of the dry mix composition on the water demand, a three-level, close to D-optimal, experimental design Ha_5 was implemented [22]. The experimental planning conditions used earlier in studying the influence of the composition of dry mixtures on the properties of hardened mortars (Table 3.18) have been supplemented by taking into account the influence factor of workability mixtures upon cone immersion (X_5) (Table 3.23).

Table 3.23 Conditions for planning experiments when studying the influence of dry mix compositions on their water consumption.

Factors		*Levels of Variation*			*Variation Interval*
Natural View	*Coded View*	*– 1*	*0*	*+1*	
Cement consumption, (C), kg/t	X_1	100	150	200	50
Slag consumption, (S), kg/t	X_2	50	75	100	25
Dust consumption, (D), kg/t	X_3	50	75	100	25
Consumption of superplasticizer, (SP-1),% of the weight of the binder	X_4	0	0,5	1	0,5
Workability of mortar mixtures by cone immersion	X_5	4	8	12	4

As a result of statistical processing of experimental data, a quadratic polynomial model of water consumption of the mortar mixture was obtained, adequate at a 95% confidence probability:

$$\begin{aligned} W = {} & 226 + 1{,}5X_1 + X_2 + 5{,}78X_3 - 24{,}2X_4 + 8{,}34X_5 \\ & + 0{,}292X_1^2 + 2{,}79X_2^2 + 1{,}79X_3^2 + 11{,}29X_4^2 + 0{,}79X_5^2 \\ & + 0{,}31X_1X_2 - 0{,}19X_1X_3 + 1{,}31X_1X_4 - 0{,}69X_1X_5 + 0{,}81X_2X_4 \\ & + 2{,}56X_2X_5 - 4{,}94X_3X_4 + 3{,}1X_3X_5 - 2{,}94X_4X_5 \end{aligned} \tag{3.31}$$

The model allows solving two types of problems:

1. predicting the change in water consumption for a given workability of the mortar mixture depending on the composition of the dry mixture.
2. determining the required water consumption for a dry mixture of a given composition.

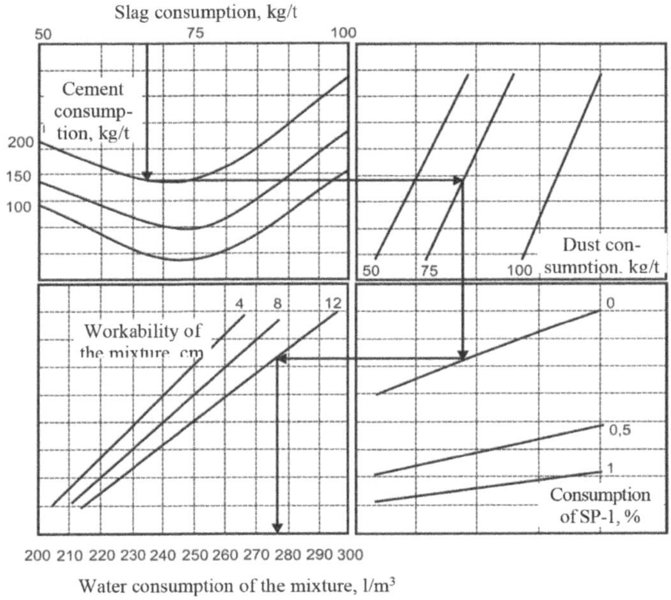

Fig. 3.22 Relationship between composition, workability of mortar mixture and its water consumption.

Solving these problems is possible using interpolation calculations by a model or a nomogram (Fig. 3.22).

Analysis of the material composition of the dry and, accordingly, the mortar mixture for its water consumption (Fig. 3.22) allows us to assume that the most significant factors influencing the required water consumption are the consumption of the superplasticizer additive and the content of dust. If the addition of a superplasticizer at a dosage of up to 1% of the mass of the filled binder (3.5...5 kg/m^3) allows for a significant reduction in water content within 20...25%, then the addition of a dust in the composition of the binder and mortar significantly increases the water content. At the same time, in the presence of the SP-1 additive, the increase in the dust content in the binder composition has a smaller effect on the increase in water content than without the addition of a superplasticizer. The water-reducing effect increases most noticeably with an increase in the SP-1 content from 0 to 0.5%. With a change in the SP-1 content from 0.5 to 1%, the increase in the water-reducing effect slows down.

Changing the content of Portland cement and slag in the composition of the binder and mortar in the studied area of variation affects water consumption to a much lesser extent.

The obtained data are in good agreement with the experimental results of other authors [75, 76]. In particular, the greatest dilution effect of the SP-1 additive in mortar mixtures on a composite binder with high initial workability is consistent with the well-known statement that the greater the volume of water in the intergranular space of cement, the greater the water-reducing effect when a superplasticizer is introduced. The fluidity of the mortar, as well as the concrete mixture, is directly dependent on the volume and fluidity of the cement paste.

Considering the chemical and mineralogical composition of clinker kiln dust, it can be classified as mineral additive with relatively low hydraulic activity. According to V.G. Batrakov [77, 78], the greatest reduction in water consumption of concrete mixtures (23...29%) can be achieved with medium-aluminate cements or cement containing mineral additives with low hydraulic activity. To obtain such an effect, a relatively small dosage of SP-1 (0.5...0.7%) is required.

The water-reducing effect of superplasticizers such as SP-1 depends on the nature of the adsorbent. In the fly clinker kiln dust as a component of the composite binder, the carbonate component prevails, which is characterized by high adsorption capacity [76].

To clarify the impact of changes in alkali and free lime content on water consumption, a study was conducted using dry mixes on a composite binder (Table 3.24). The content of Na_2O and CaO_{free} in the dust was adjusted by introducing Na_2CO_3 and quicklime additives.

Table 3.24 The influence of Na2O and CaOfree content in dust on the water consumption of mortar mixtures.

No	*Binder Composition, kg/t of dry mix*				*Na_2O and CaO_{free} Content, %*		*Workability, cm*	*Water Consumption, l/m^3*
	C	*Sl*	*D_1*	*SP-1*	*Na_2O*	*CaO_{free}*		
1	200	100	100	4	2,5	2,8	12	235
2					5,5	2,8	12	228
3					2,5	5,8	12	249
4					5,5	5,8	12	245
5	100	100	100	3	2,5	2,8	4	202
6					5,5	2,8	4	201
7					2,5	5,8	4	217
8					5,5	5,8	4	215
9	100	50	50	2	2,5	2,8	4	195
10					5,5	2,8	4	190
11					2,5	5,8	4	201
12					5,5	5,8	4	200

The studies were conducted on mortars of two characteristic compositions, with the workability of the mortar mixtures of 4 and 12 cm. As shown by the obtained data (Table 3.24), increasing the Na_2O content in the dust from 2.5 to 5.5% did not change or slightly reduced the water consumption. The increase in the content of free lime in the dust composition, as expected, almost regardless of the Na_2O content, caused a noticeable increase in the water consumption of mortar mixtures.

The water-holding capacity was determined by testing mortar mixtures 12 cm thick, placed on blotting paper after 10 minutes of exposure and calculated the using the formula:

$$B_{\text{hold}} = \frac{m_1 - m_2}{m_1} \cdot 100, \quad (3.32)$$

where, m_1 – mass of the water in the mixture;

m_2 – mass loss of the water, *g*.

The results of determining B_{hold} for mortar mixtures based on the characteristic compositions of dry mixtures, taking into account possible fluctuations in R_2O and CaO_{free}, are given in Table 3.25. The water-holding capacity was insufficient for mortars of compositions 9…12, which are characterized by a relatively small volume of cement paste.

Table 3.25 Results of determination of B_{hold} for mortar mixtures.

No	*Binder Composition, kg/t of Dry Mix*				*Composition of Na_2O and CaO_{free}, %*		*Normal Consistency, %*	*Water-holding Capacity, %*
	C	*S*	*D_1*	*SP-1*	*Na_2O*	*CaO_{free}*		
1	200	100	100	4	2,5	2,8	31,9	96,4
2					5,5	2,8	30,8	95,5
3					2,5	5,8	33,2	97,2
4					5,5	5,8	32,5	97,1
5	100	100	100	3	2,5	2,8	31,4	95,3
6					5,5	2,8	30,5	95,2
7					2,5	5,8	34,2	96,8
8					5,5	5,8	33,3	95,6
9	100	50	50	2	2,5	2,8	30,8	91,2
10					5,5	2,8	29,6	90,3
11					2,5	5,8	33,5	93,4
12					5,5	5,8	33,1	93,5

Note: 1. Mortar mixtures of compositions 1…4 had a workability of 12 cm, 5…12 – 4 cm. **2.** The yield coefficient of the mortar mixture of compositions 1…4 was on average 0.7, 5…12 – 0.6.

The introduction of dust into the composition of composite binders, increasing the viscosity of pastes based on them, contributes to an increase in the normal consistency and water-holding capacity of mortars. An increase in the composition of binder alkalis, which are contained in the dust no more than 2.5...5.5%, has practically no effect on the water-holding capacity of mortars, and free lime within 2.8...5.8% contributes to some increase in it. Calculations (Table 3.26) show that for all compositions of mortars with a water content less than $1.35K_{n.c.}$, ($K_{n.c.}$ is the normal consistency of the binder in unit parts) the water-holding capacity is provided at least 95%. If this requirement is not met, it becomes less than the required value.

Table 3.26 Water-holding capacity of mortar mixtures and water consumption values.

Composition No. according to Table 3.25	*Binder Consumption, kg/t (C+Sl+D₁)*	*Output Ratio β*	*Binder Consumption, kg/m³ of mortar* $\frac{(C+Sl+D_1)}{\beta}$	*Water Content, l/m³*	*Water-holding Capacity, %*
1	400	0,7	571	235	96,4
2	400	0,7	571	228	95,5
3	400	0,7	571	249	97,2
4	400	0,7	571	245	97,1
5	300	0,6	500	2022	95,5
6	300	0,6	500	201	95,2
7	300	0,6	500	217	96,8
8	300	0,6	500	215	95,6
9	200	0,6	333	195	91,2
10	200	0,6	333	190	90,3
11	200	0,6	333	201	93,4
12	200	0,6	333	200	93,5

The suitability time of mortar mixtures should, obviously, be determined by the time during which the workability decreases to the minimum permissible. The change in the workability of mixtures over time should obviously be affected by all factors that determine the rate of cement hydration. In addition to W/C and temperature, such factors include activity, dispersion of the binder, its setting time, etc.

We studied the change in the workability of the mortar mixture of two characteristic compositions with the initial immersion of the reference cone of 14 and 8 cm and different contents in the binders alkali and free lime. The experiments were conducted at a mortar mixture temperature of $20 \pm 3°C$ with measurement of the workability of the mixture immediately after mixing and then every 30 minutes.

Portland cement was used, differing in setting time: No. 1 with the initial setting time 2 hours 15 minutes, and No. 2 setting time 1 hour 30 minutes.

The results of the experiments are shown in Figs. 3.23, 3.24. They show that the nature of the curves of the decrease in the workability of mortar mixtures over time is satisfactorily approximated by a function of the type:

$$y = a^x, \tag{3.33}$$

where, x – power index characterizing the duration of the mixture's aging;

a – the base of the degree, the value of which is influenced by the composition of the dry and, accordingly, the mortar mixture, the content of alkalis and free lime.

A faster decrease in workability is facilitated by an increase in the content of the binder component of the mortar mixture and an acceleration of the setting time of cement as its main component.

Increasing the duration of the initial setting time, as well as reducing the binder content, expands the range of possible alkali content in dust.

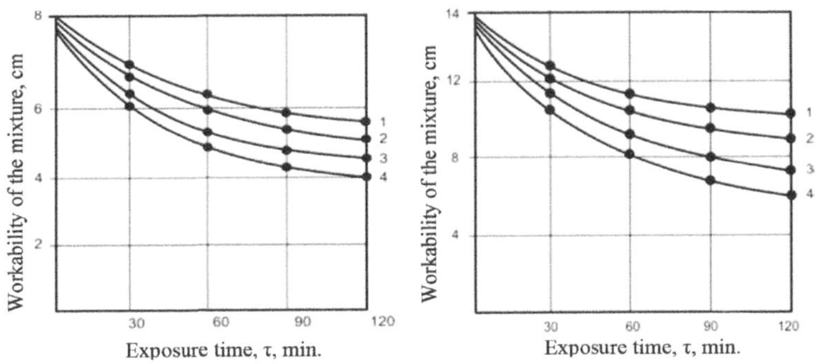

Fig. 3.23 The influence of the holding time on the workability of the mortar mixtures (composition of the dry mixture: C = 100 kg/t, S = 100 kg/t, D_1 = 100 kg/t, initial of setting time 2 hours 15 minutes): 1 – R_2O = 2,5%, CaO_{free} = 5,8% ; 2 – R_2O = 2,5%, CaO_{free} = 2,8%; 3 – R_2O = 5,5%, CaO_{free} = 2,8%; 4 – R_2O = 5,5%, CaO_{free} = 5,8%.

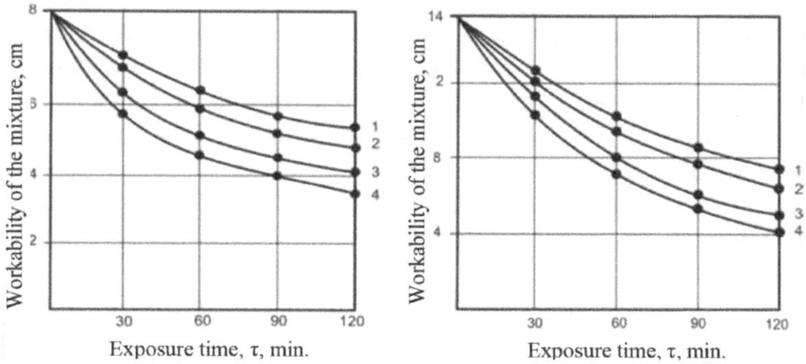

Fig. 3.24 The influence of the holding time on the workability of the mortar mixture (composition of the dry mixture: C = 200 kg/t, S = 100 kg/t, D_1 = 100 kg/t, Sd = 600 kg/t; start of hardening 1 hour 30 minutes):1 – R_2O = 2,5%, CaO_{free} = 5,8%; 2 – R_2O = 2,5%, CaO_{free} = 2,8%; 3 – R_2O = 5,5%, CaO_{free} = 2,8%; 4 – R_2O = 5,5%, CaO_{free} = 5,8%.

Thus, the conducted studies show the possibility of using cement-dust-slag binders for the production of dry building mixtures for masonry and mortars at 3D printing. Dry mixtures obtained by mixing the initial components of the binder, plasticizer additive and fine-grained quartz sand are characterized by a 28-day compressive strength of 10–30 MPa with a workability of the mortar mixture of 8–10 cm. By the 7-day age, the strength of the mortars reaches 70% of the brand value.

3.7 Porous Mixtures on Cement-slag Binder with Ash and Hardening Accelerator Additives

To increase the thermal resistance of structures during the construction of buildings and structures, including using additive technology, it is advisable to use porous lightweight concrete. Porous concretes based on dry mixes, if their properties meet the necessary requirements, can be successfully used in 3D concreting of wall structures. The effect of using such concretes is achieved if its thermal conductivity is no less than other wall materials. Along with reducing heat consumption, such concretes must provide high adhesion between layers, water, and frost resistance.

In the studies, UFAPORE CC85, produced by Bang & Bonsomer of the Mapei concern, was used as a concrete pore-forming additive, which is a mixture of anionic surfactants. In addition to plasticizing and increasing the frost resistance of concrete, the additive slightly reduces water separation and sedimentation stratification of mixtures.

During the research, the binder was produced by mixing commercial Portland cement CEM I - 42,5N and ground granulated blast furnace slag in a ratio of 1:1. Fly ash or quartz sand with M_f= 1.9 were additionally introduced into the mortar mixtures. To increase structural strength, Na_2SO_4 was added in an amount of 1% by weight of the binder. The water consumption was varied to achieve the required workability of the cone immersion 8…10 cm.

For the mortar mixture, workability and extrusion, structural strength were determined, and for hardened mortars, the average density and compressive strength of cube samples with an edge of 70.7 mm, which hardened in air-dry conditions at the age of 3 and 28 days. The main experimental results of the research are given in Table 3.26 and Figs. 3.25…3.27.

Analyzing the obtained data, it can be noted that the water requirement of mixtures to achieve the required workability significantly decreases with increasing the amount of pore-forming additive, but up to a certain limit, which depends on the average density and the ratio of "”binder:aggregate”. With the same content of blowing agent, reducing the ratio of "binder:aggregate" from 1:1 to 1:2 slightly reduces the water consumption of a mixture in which the aggregate is fly ash. Further reducing this ratio to 1:3 significantly reduces water consumption and water-binding ratio.

The change in the structural strength of the studied compositions of mortars is shown in Figure 3.25.

Analyzing the change in structural strength after 20 minutes of porous mortars, it can be stated that its maximum values are found in compositions using quartz sand as an aggregate, regardless of the consumption of the porous agent. After mixing, its maximum values are characteristic of compositions No. 1, 7, 10, 11 (Table 3.26 and Fig. 3.25).

The strength of the investigated mortars, as expected, depends on the porosity of the material, which is influenced by the content of the pore former and which can be estimated by the average density (Fig. 3.26).

The fly ash-based mixture is characterized by an average density of 1200...1400 kg/m^3 and a strength of 13...20 MPa. At the same time, an increase in density from 1200 to 1400 kg/m^3 is accompanied by an increase in strength by 20...25%, for a mixture with ratio of "binder:ash"=1:1, a further increase in average density does not lead to a significant increase in strength. The same dependence is observed when using a quartz sand mixture as an aggregate; in this case, mortars (sand concretes) with a strength of 15…25 MPa are obtained. For them, the average density increases to a density of 1300...1600 kg/m^3. Light mortars with the composition Binder:Sand=1:2 with an increase in density from 1200 to 1450 kg/m^3 are characterized by a linear increase in strength by 60…70%, the maximum value of compressive strength is 20…25 MPa.

Table 3.27 Properties of porous mortars for 3D concreting.

No. in Order	*Ash (A) Sand (S)*	*Binder: A*	*Binder: S*	*Consumption of Pore Former, %*	*Den-sity ρ_0, kg/m^3*	*Structural Strength through, KPa*		*Compressive Strength at Age, MPa*	
						20 min	*1 hour*	*3 days*	*28 days*
1	Fly ash	1:1	-	0,08	1290	4260	9370	9,7	17,8
2		1:1	-	0,11	1170	3960	8315	5,2	14,3
3		1:2	-	0,04	1300	4250	8925	4,7	18,8
4		1:2	-	0,08	1250	4650	8740	4	12,3
5		1:2	-	0,11	1210	4450	8010	3,7	8,8
6		1:3	-	0,04	1390	4550	8190	8,2	13,2
7	Sand	-	1:1	0,08	1470	4650	9300	7,1	20,1
8		-	1:1	0,11	1310	4500	9000	10,5	17,6
9		-	1:2	0,04	1520	4850	8730	5,6	24,8
10		-	1:2	0,08	1450	4880	9175	10,5	17,0
11		-	1:3	0,04	1600	5100	9585	7,7	15,3

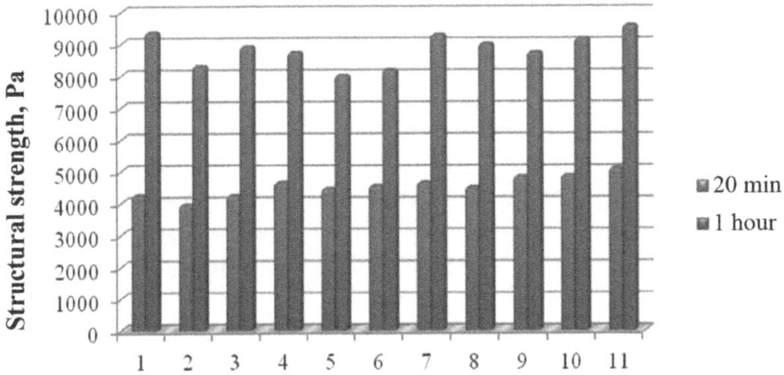

Fig. 3.25 Change in structural strength of porous mortars after 20 min and 1 h, (composition numbers according to Table 3.25).

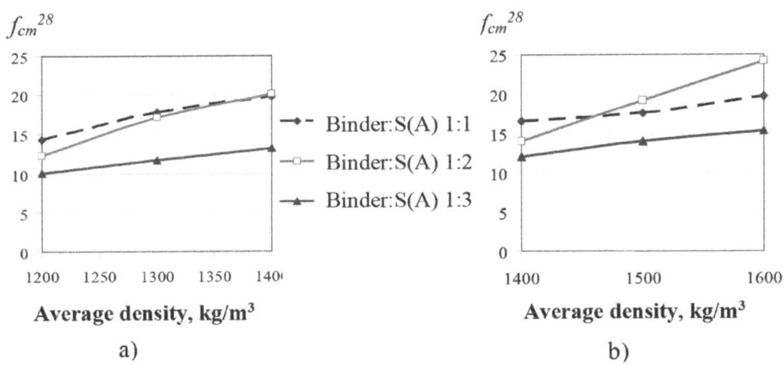

Fig. 3.26 The influence of average density on the strength of porous mortars at 28 days of age (a) – aggregate – quartz sand (S); (b) – filler – fly ash (A)).

Increasing the content of the pore-forming agent (Fig. 3.28) reduces the density of the concrete, however, after a certain limit (approximately 0.1% of the mass of the mixture) the effect becomes unstable. Further reduction of the density when using the adopted pore-forming agent is impossible. Obviously, additives with more intense foaming should be used, although their negative impact on the hardening kinetics and strength of concrete can be expected.

Based on the obtained experimental data (Table 3.28), the compositions of dry building mixtures for porous mortars were proposed (Table 3.28).

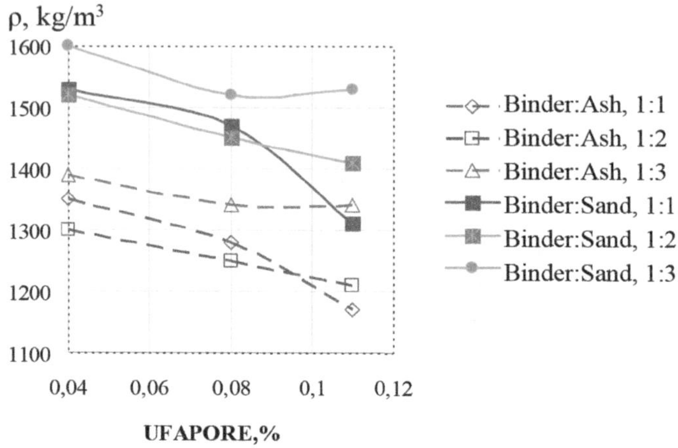

Fig. 3.27 Influence of the content of the UFAPORE CC85 pore former and the filler on the average density of the porous mortars.

Table 3.28 Recommended compositions of porous mortars for 3D concreting.

No. in Order	*Compressive Strength, MPa*	*Average Density ρ_0, kg/m³*	*Type of Filler*	*Consumption, kg, per 1 t of Dry Building Mixes*			*Water Consumption during Mixing, l*
				Binding	*Sand, Ash*	*UFAPORE*	
1	20	1500	Sand	285	715	0,4…0,5	100…120
2		1300	Ash	350	650	0,4…0,5	140…150
3	15	1350	Sand	310	690	0,9	110...130
4		1250	Ash	300	700	1,1	160...170
5	10	1200	Ash	260	740	0,9...1,0	130...150

4

Mixtures Based on Gypsum-Cement-Pozzolan Binders

4.1 Gypsum-containing Mixtures for 3D Printing

The most famous representative of quick-hardening binders produced by the construction industry are gypsum binders. However, the main problem of these binders is the low water resistance of the hardened stone, which significantly limits the scope of its use. One of the most effective solutions in this direction can be considered the development of gypsum-cement pozzolanic (GCPB) and gypsum-cement-slag binders (GCSlB) [21]. The main idea of the development is to obtain a product that would combine the best properties of building gypsum (ability to quickly harden and gain strength) and Portland cement (water resistance and strength). These advantages allow the use of GCPD and GCSlB as binders for the production of concrete and mortars for construction of 3D printers.

At the current stage of development of the construction industry, a large number of new generation superplasticizers have been developed. Below are the results of comparative studies of the effectiveness of plasticizing additives of various types on the complex properties of mixtures based on compositional gypsum binders intended for 3D printing.

Research into the effects of plasticizer additives. To maximize the reduction of water consumption and the corresponding increase in strength, plasticizing additives of various types were added to two typical compositions of gypsum-cement-slag binder (GCSlB) mixed with quartz sand ($M_f = 1.9$) in a ratio of 1:2 (by weight) in an amount of 0.5% (in terms of dry matter) of the binder mass. To obtain the binder, gypsum G-5, Portland cement CEMI-42.5, and blast furnace granulated slag were used as the initial components.

To regulate the setting time, a setting retarder – tartaric acid, was added to the mixture in an amount of 0.3% by weight of gypsum. As plasticizing additives at this stage, the following surface active additives

DOI: 10.1201/9781003685883-4

were used: technical lignosulfonate (LST), naphthalene-formaldehyde superplasticizer SP-1, polycarboxylate superplasticizers Melflux 2651F and PCE-7, as well as polyacrylate SP-3.

The main task was to determine the effect of additives on the following properties of mortar mixtures:

- extrudability (formability);
- shelf life (initial of setting time);
- structural strength through 10 minutes after mixing;
- compressive strength at 2 hours, 1, 7, and 28 days;
- bond strength between layers at 28 days.

Water consumption in the experiments was determined from the condition that the mixture achieved a given spread of 160 ± 5 mm using a Suttard viscometer. Formability was determined by the ability of the mixture to be extruded from the nozzle of a laboratory 3D printer without the appearance of cracks and delamination along the length of the beam. The shelf life was determined in minutes from the moment of mixing to the moment the initial setting time mixture, at which its further extrusion molding was no longer possible. Structural strength, after 10 minutes, was determined by the specific maximum load that the mixture could withstand without deformation and failure. The bond strength between the layers was determined by the value of the tensile strength at splitting at the layer boundary. The resulting mixtures were tested for strength after exposure to air for two hours, after 1, 7, and 28 days of hardening under normal conditions ($t = 20°C$, relative humidity $\varphi = 90...100\%$).

The results of experimental studies of the influence of plasticizing additives on the strength characteristics of mortar mixtures are given in Table 4.1.

The obtained results of the initial setting time (Fig. 4.1) and structural strength (Fig. 4.2) of mixtures based on gypsum-cement-slag binder when using additives of different types allow us to conclude that the superplasticizers LST and SP-1 in the composition with the addition of tartaric acid additionally increase the initial setting in both compositions of GCSlB.

All plasticizers, reducing the water-binding ratio, increase structural strength.

The results obtained (Figs. 4.3...4.7) indicate that the greatest strength of gypsum-cement-slag binder, both at the age of 2 hours and at later hardening times, when using all the above-mentioned plasticizing additives, is ensured with the following component composition: building gypsum – 50%, Portland cement – 20%, blast furnace granulated slag – 30%. It was also found that the most effective plasticizing additive, for

Table 4.1 Results of experimental studies of the influence of plasticizing additives on the initial of setting time and strength characteristics of the mortars based on GCSIB.

Composition of the GCSIB, %	*Type of Additive*	*W/Bind*	*Initial Setting Time, min.*	*Structural Strength, after 10 min, Pa*	*Compressive Strength at Age, MPa*				*Bond Strength between Layers, MPa*
					2 Hours	*1 Day*	*7 Days*	*28 Days*	
G-60 C-15 Sl-25	-	0,63	28	3640	3,8	4,6	5,8	6,6	3,2
	LST	0,55	31	3850	3,7	5,5	8,7	11,1	3,2
	Mel-flux 2651F	0,42	27	3880	8	10,3	13,2	16,1	4,7
	SP -1	0,50	30	4420	4,6	6,7	12,3	15,8	4,2
	SP-3	0,48	28	4200	5,1	10,2	12,3	13,9	4,3
	PCE-7	0,41	28	4050	5,5	10,2	13,4	14,1	4,6
G-50 C-20 Sl-30	-	0,62	26	3370	4,5	4,6	6,4	7	3,8
	LST	0,56	30	3570	4,1	5,7	9,1	12	3,6
	Mel-flux 2651F	0,40	25	3610	11,4	17,2	17,6	20,2	5,5
	SP-1	0,50	28	4080	4,5	6,9	12,6	16,2	4,1
	SP-3	0,45	26	3880	5,6	10,6	12,5	14,4	4,7
	PCE-7	0,42	25	3750	5,7	10,5	13,6	14,3	4,8

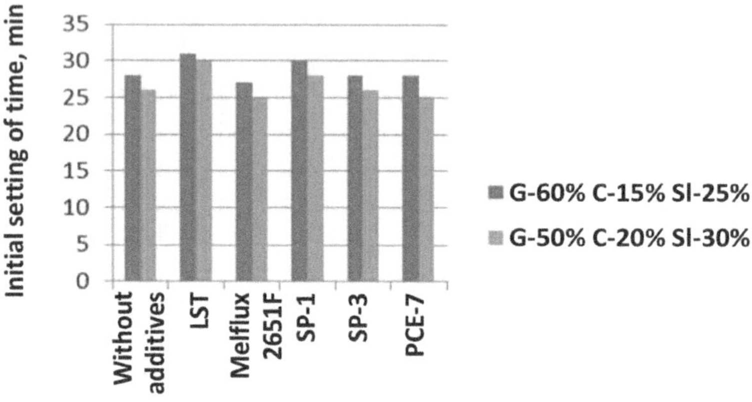

Fig. 4.1 The influence of plasticizing additives on the initial setting time of mixtures.

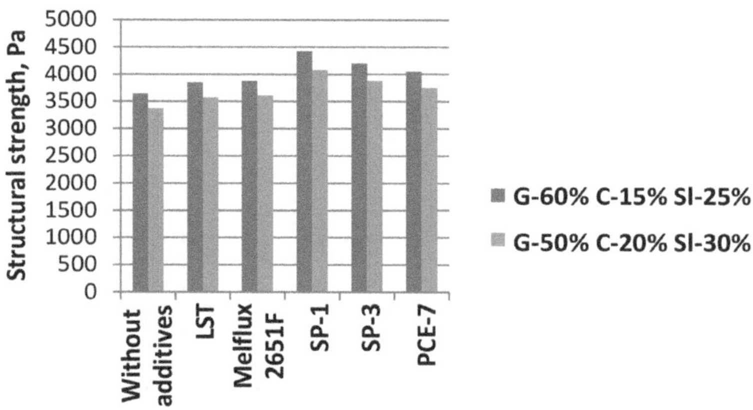

Fig. 4.2 The effect of plasticizing additives on the structural strength of mortars, after 10 min.

this component composition of the binder, is the polycarboxylate-based additive Melflux 2651 F. At the same time, the strength of the mortar after the second hour of hardening is 11.4 MPa and 20.2 MPa after 28 days. The increase in the strength of the samples when using the superplasticizer additive Melflux 2651 F is explained by its stronger water-reducing effect compared to other additives used.

The obtained results of the bond strength between the layers of mortars based on GCSlB at the age of 28 days (Fig. 4.7) indicate that the

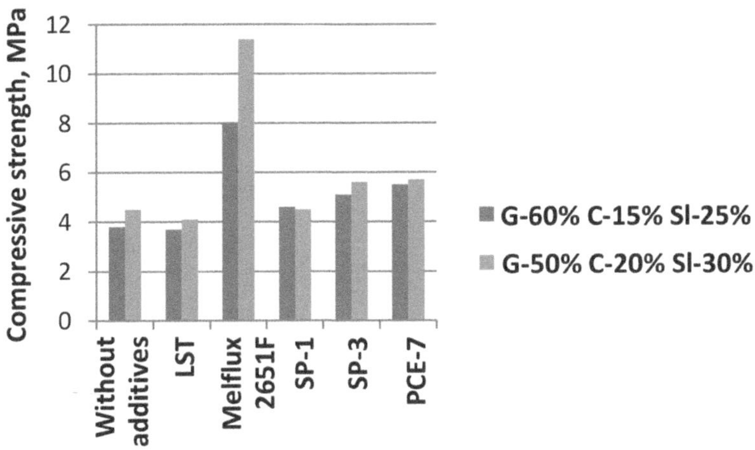

Fig. 4.3 The effect of plasticizing additives on the compressive strength of mortars at the age of 2 hours.

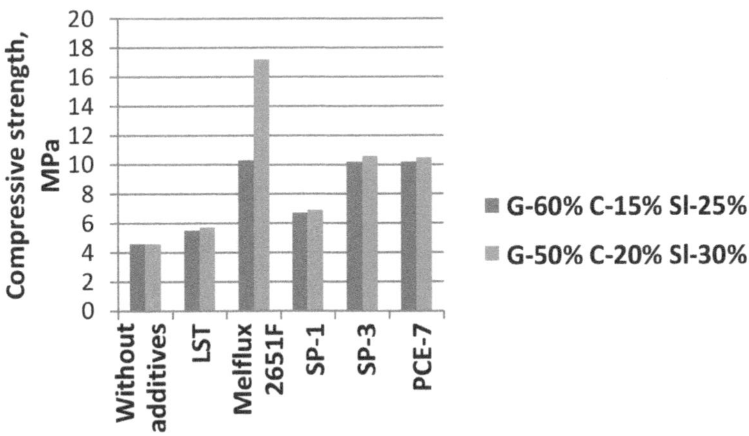

Fig. 4.4 The effect of plasticizing additives on the compressive strength of mortars at the age of 1 day.

greatest strength is also provided when using the Melflux 2651 F additive. The water-reducing effect of the used Melflux 2651 F additive reaches at least 30%, which allows to significantly reduce the water/binder ratio and increase the strength. The splitting strength, which determines the adhesion strength of extrusion layers, reaches 5.4 MPa at 28 days.

When mixing polycarboxylate-type superplasticizers with Portland cement, the polycarboxylate ester is modified by replacing Na^+ ions in

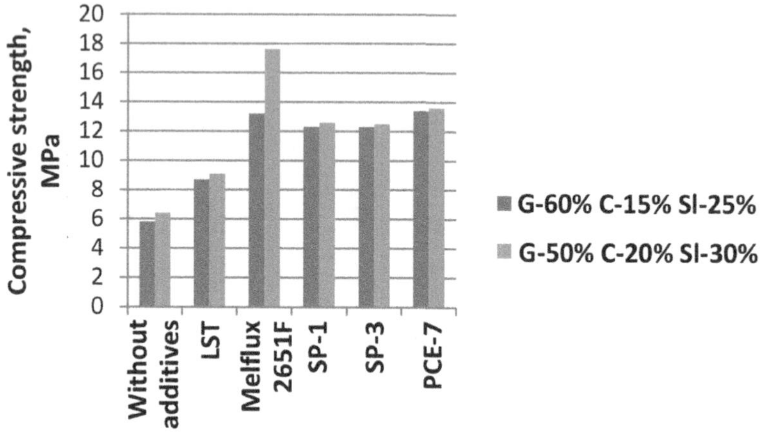

Fig. 4.5 The effect of plasticizing additives on the compressive strength of mortars at the age of 7 days.

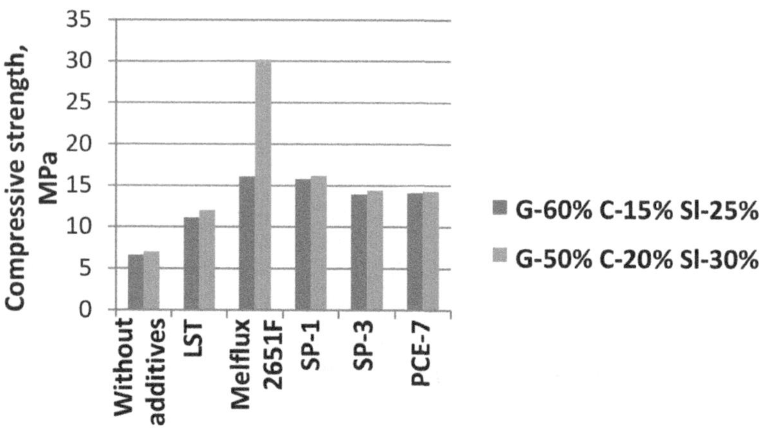

Fig. 4.6 The effect of plasticizing additives on the compressive strength of mortars at the age of 28 days.

the active radicals of this compound with Ca^+ ions. In the absence of an alkaline environment in the gypsum mortar, which occurs due to the formation of calcium hydroxide during the hydration of cement, the above modification is impossible. The introduction of gypsum binder, Portland cement, granulated blast furnace slag, and a polycarboxylate-type superplasticizer into the composition provides an intensive water-reducing effect due to the combined action of both the electrostatic and steric (spatial) effects of the modified polycarboxylate ester molecules.

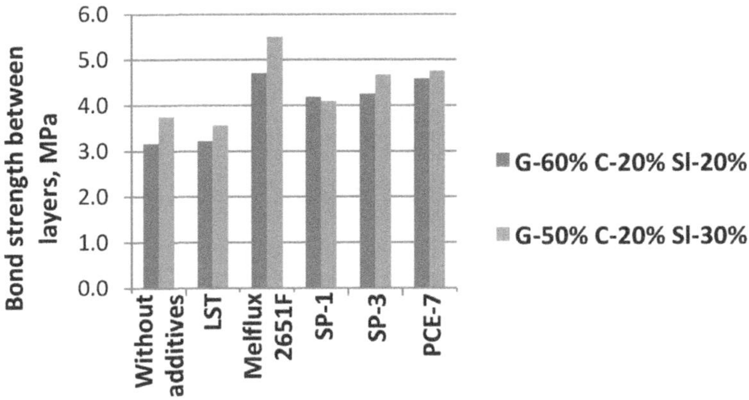

Fig. 4.7 The influence of plasticizing additives on the bond strength between layers of mortars at the age of 28 days.

4.2 Gypsum-containing Porous Fiber-reinforced Mortars

One of the important components in mixtures for 3D printing is fiber. It is introduced into dry mixes, concrete and mortars, acting as a micro-reinforcing component, modifying the structure of building conglomerates at the micro- and macro-levels. This increases strength, especially bending strength, tensile strength during splitting, increases crack resistance, thermal stability, shortens construction times, and reduces materials consumption.

The use of fiber allows, in certain cases, to provide the necessary load-bearing capacity of structures without the use of bar reinforcement, provided that high structural strength of the freshly formed layer of the structure is ensured at a given workability of the mixture. Table 4.2 lists the types of fiber used as components of mixtures for 3D printers[10].

Table 4.2 Types of fibers in 3D printer mixtures.

Fiber Type	*Advantages*
Glass	• Increases impact resistance; • Achieves high density and uniformity of reinforcement
Polypropylene	• Reduces the specific gravity of the mixture; • Increases the wear resistance and abrasion resistance of structures; • Increases water resistance and frost resistance

Fiber Type	*Advantages*
Basalt	• Has electrical insulating properties; • Does not support combustion; • Environmentally friendly
Steel	• Reduces the thickness and weight of load-bearing structures; • Increases resistance to dynamic loads;

To reduce the average density and correspondingly increase the thermal resistance of the enclosing structures, the pore-forming additive (UFAPORE CC85) was additionally introduced into the composition of the mixtures based on the proposed GCShB. To reduce the thermal conductivity of mortars, light aggregates (for example, expanded perlite or vermiculite sand) can also be added to their composition; however, the introduction of pore-forming additives is usually a more appropriate method for both economic and technological reasons.

To obtain a homogeneous binder with increased strength, it was produced by co-grinding construction gypsum plaster, Portland cement, granulated blast furnace slag, and powdered polycarboxylate superplasticizer. The content of gypsum plaster of the G-10 brand in the binder was 50%, Portland cement CEM I – 20%, blast furnace granulated slag – 30%. To regulate the setting time, a setting retarder was introduced into the mixture – tartaric acid, in an amount of 0.3% of the gypsum mass. The specific surface area of the binder reached 350 m^2/kg, it was characterized by a strength of 16.8 MPa at the age of 2 hours and 35 MPa at the age of 28 days, which was 1.5 times higher than the strength of the binder made by mixing the individual components. The studies used polypropylene fiber 20 mm long.

To test the possibility of using this type of binder for the preparation of fast-hardening mortar mixtures for a 3D construction printer, beam samples measuring 40 × 40 ×160 mm were manufactured with a binder:sand ratio of 1:1. Water consumption in the experiments was determined from the condition of ensuring a given spread of 160 ± 5 mm according to the Suttard viscometer, which ensured the necessary formability of the mixture. Formability was determined by the ability of the mixture to be extruded from the 3D printer nozzle without the appearance of cracks and delamination along the length of the beam. Structural strength, 10 minutes after mixing, was determined by the specific load on the sample. The resulting mixtures were tested for strength after exposure to air for two hours and 28 days of hardening under normal conditions ($t = 20°C$, $\varphi = 90...100\%$).

All experimental studies at this stage were performed using mathematical experimental planning. For this purpose, a three-level three-factor plan B3 was implemented [22]. The experimental design conditions are given in Table 4.3.

The planning matrix and the experimental results obtained are given in Table 4.4.

Table 4.3 Conditions for planning an experiment.

Number	*Factors*		*Levels of Variation*			*Variation Interval*
	Coded View	*Natural View*	*– 1*	*0*	*+1*	
1	X_1	Fiber consumption (F), kg/m^3	0	2,5	5	2,5
2	X_2	Porizer consumption (P), %	0	0,2	0,4	0,2
4	X_3	Water binding ratio (W/Bind)	0,45	0,50	0,55	0,02

Analyzing the obtained results, it can be concluded that the influence of the water-binding ratio and the pore-forming agent on the initial setting time is quite insignificant which varies within narrow limits from 25 to 29 min. Structural strength mainly depends on the fiber consumption, with an increase in its content, the strength increases by 8...10%. The introduction of a pore-forming agent and an increase in the W/Bind ratio slightly reduces the structural strength.

As a result of the analysis of the obtained experimental and statistical models the graphical dependencies built on their basis (Figs. 4.8, 4.9), it was established that the most significant factor affecting the average density of mortars made on the basis of GCSIB is the consumption of the pore-forming additive. Increasing its content to 0.4% of the binder mass allows reducing the average density of the samples by an average of 20%. At the same time, the average density of the samples decreases from 2000–2100 kg/m^3 to 1600–1700 kg/m^3, which accordingly leads to a significant decrease in thermal conductivity. Also, an increase in the water-binding ratio leads to a slight decrease in the average density of the samples, which accordingly causes an increase in the total porosity of the hardened mortar.

The nature of the influence of varied factors on the strength characteristics of mortars for 3D concreting as follows from the analysis of the complex of obtained experimental statistical models (Table 4.5),

Table 4.4 Results of experimental studies.

Number	*Fiber Content kg/m³*	*Porizer Content,%*	*W/Bind*	*Initial of Setting Time, min*	*Structural Strength after 10 min, Pa*	*Splitting Tensile Strength, MPa, at Age*		*Comp-ressive Strength, MPa, at Age*		ρ_0, *kg/m³*
	X_1	X_2				*2 hours*	*28 days*	*2 hours*	*28 days*	
1	5	0,4	0,55	29	4030	3,75	9,76	6,0	13,0	2049
2	5	0,4	0,45	27	4350	4,15	11,71	8,2	30,6	1949
3	5	0	0,55	27	5240	4,74	10,39	10,3	35,4	1735
4	5	0	0,45	25	5410	5,99	12,64	11,1	49,0	1670
5	0	0,4	0,55	29	3650	3,23	7,42	5,9	17,0	2115
6	0	0,4	0,45	27	3835	3,58	8,25	7,5	27,4	2027
7	0	0	0,55	27	4050	3,71	9,05	8,6	26,6	1637
8	0	0	0,45	25	4300	4,91	10,18	8,8	33,0	1584
9	5	0,2	0,50	27	4410	4,12	11,11	9,1	28,0	1801
10	0	0,2	0,50	27	4000	3,32	8,71	7,9	22,0	1791
11	2,5	0,4	0,50	28	3880	3,74	8,20	7,3	17,4	2042
12	2,5	0	0,50	26	4320	4,90	9,48	10,1	31,4	1664
13	2,5	0,2	0,55	28	4020	3,66	8,02	8,5	19,5	1827
14	2,5	0,2	0,45	26	4320	4,46	9,56	9,7	31,5	1751
15	2,5	0,2	0,50	27	4200	3,92	8,81	9,0	23,0	1796
16	2,5	0,2	0,50	27	4250	3,92	8,76	9,2	23,1	1794
17	2,5	0,2	0,50	28	4230	3,91	8,9	9,1	23,0	1800

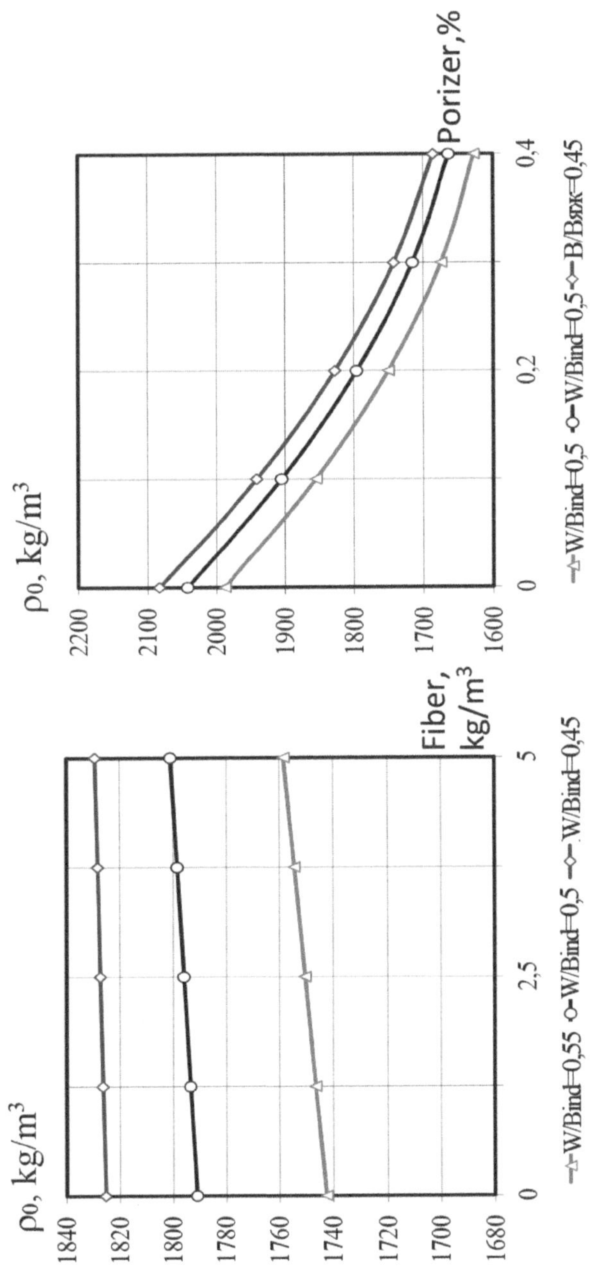

Fig. 4.8 Influence of varied factors on the average density of mortars for 3D concreting made on the basis of GCSIB.

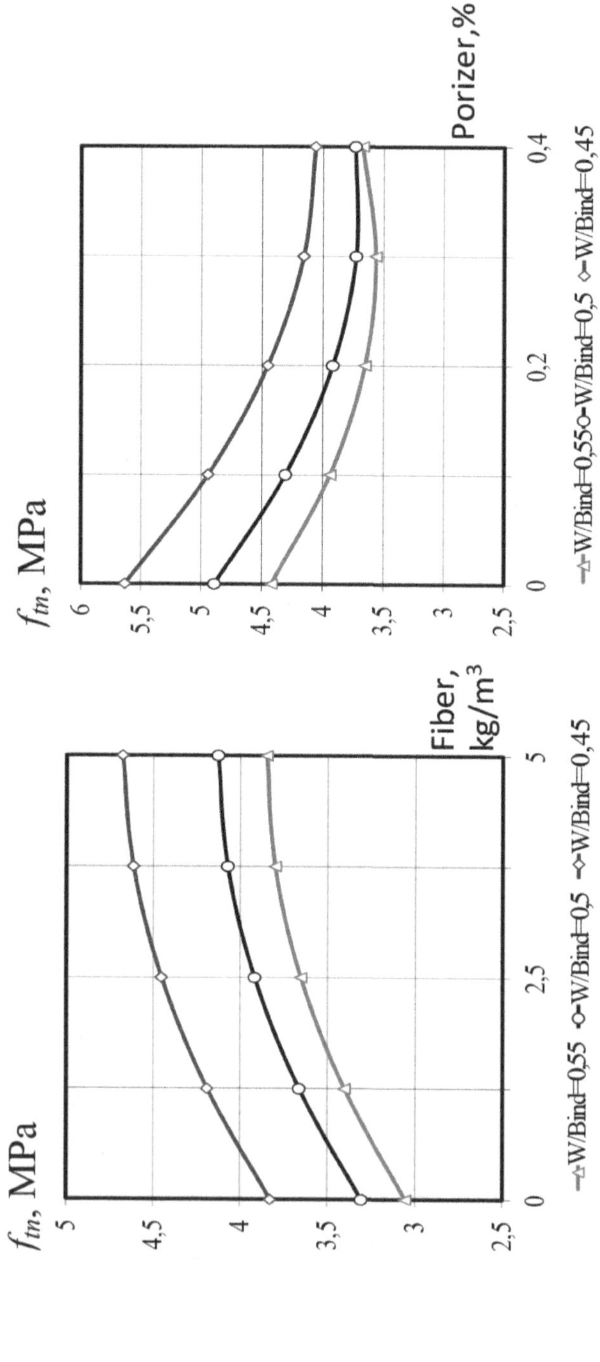

Fig. 4.9 Influence of variable factors on the tensile strength during splitting of mortars for 3D concreting made on the GCSIB at the age of 2 hours.

does not change significantly with increasing curing duration. The decrease in the strength of mortar, both in compression and in tension during splitting, is caused by an increase in the water-binding ratio and an increase in the content of the pore-forming additive, which is mainly due to a sharp increase in the porosity of the samples. This effect can be somewhat reduced by increasing the content of polypropylene fiber. Due to the three-dimensional dispersed reinforcement of the cement matrix of the mortar with fiber, its tensile strength at splitting increases by an average of 25% compared to unreinforced samples. The increase in fiber content also has a positive effect on the compressive strength. In this case, the increase in strength is on average 15–20%.

Table 4.5 Experimental statistical models of the properties of porous mortars made on the basis of GCShB.

Output Parameters	*Models*	
Average density, kg/m^3	$\rho_0 = 1796 + 5X_1 - 189{,}2X_2 - 38{,}4X_3 + 56{,}9X_2^2 - 7{,}1X_3^2 - 41X_1X_2 + 3X_1X_3 + 9X_2X_3$	(4.1)
Structural strength after 10 min, Pa	$P_m = 4161 + 361X_1 - 358X_2 - 122X_3 + 107X_1^2 + 2X_2^2 + 72X_3^2 - 175X_1X_2 - 7X_1X_3 - 10X_2X_3$	(4.2)
Tensile strength at split at the age of 2 hours, MPa	$f_{tn}^{2hours} = 3{,}92 + 0{,}4X_1 - 0{,}58X_2 - 0{,}4X_3 - 0{,}2X_1^2 + 0{,}4X_2^2 + 0{,}14X_3^2 - 0{,}13X_1X_2 - 0{,}01X_1X_3 + 0{,}21X_2X_3$	(4.3)
Compressive strength at the age of 2 hours, MPa	$f_{cm}^{2hours} = 9 + 0{,}6X_1 - 1{,}38X_2 - 0{,}6X_3 - 0{,}5X_1^2 - 0{,}3X_2^2 + 0{,}1X_3^2 - 0{,}4X_1X_2 - 0{,}15X_1X_3 - 0{,}35X_2X_3$	(4.4)
Tensile strength at split at 28 days, MPa	$f_{tn}^{28} = 8{,}81 + 1{,}2X_1 - 0{,}64X_2 - 0{,}77X_3 + 1{,}1X_1^2 + 0{,}03X_2^2 - 0{,}016X_3^2 + 0{,}25X_1X_2 - 0{,}26X_1X_3 + 0{,}08X_2X_3$	(4.5)
Compressive strength at the age of 28 days, MPa	$f_{cm}^{28} = 23{,}0 + 3X_1 - 7X_2 - 6X_3 + 2X_1^2 + 1{,}4X_2^2 + 2{,}5X_3^2 - 3{,}2X_1X_2 - 1{,}8X_1X_3 - 1{,}03X_2X_3$	(4.6)

The results obtained indicate that based on the developed gypsum-cement-slag binders with the addition of sand, it is possible to obtain fast-hardening mortar compositions that are characterized by increased strength and thermal resistance and are not inferior to conventional ceramic bricks in terms of their physical and mechanical characteristics.

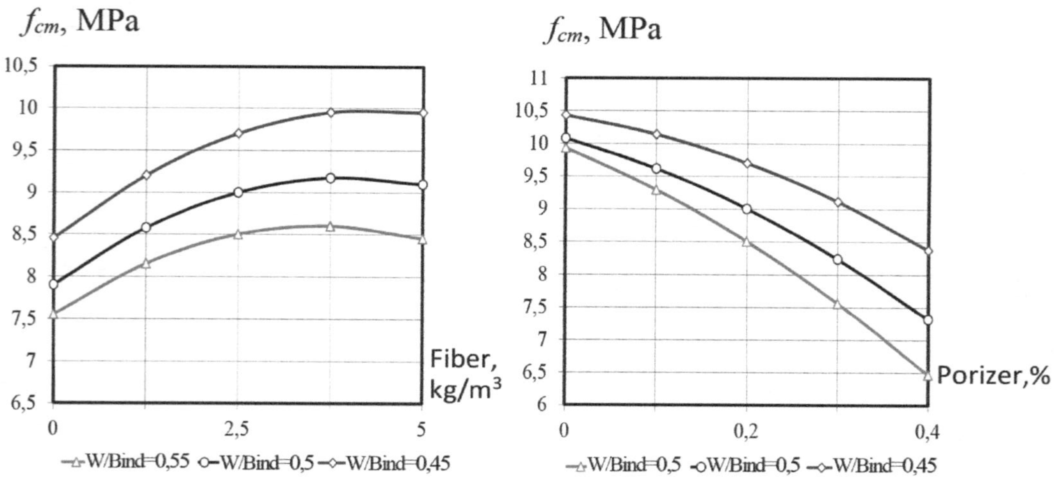

Fig. 4.10 Influence of varied factors on the compressive strength of mortars for 3D concreting made on the GCSIB at the age of 2 hours.

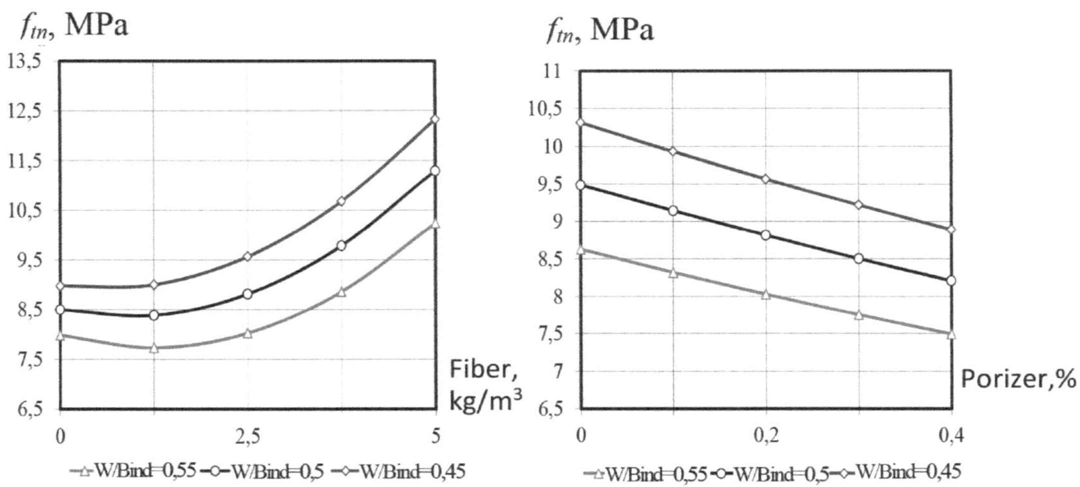

Fig. 4.11 Influence of variable factors on the tensile strength during splitting of mortars for 3D concreting made on the GCSIB at the age of 28 days.

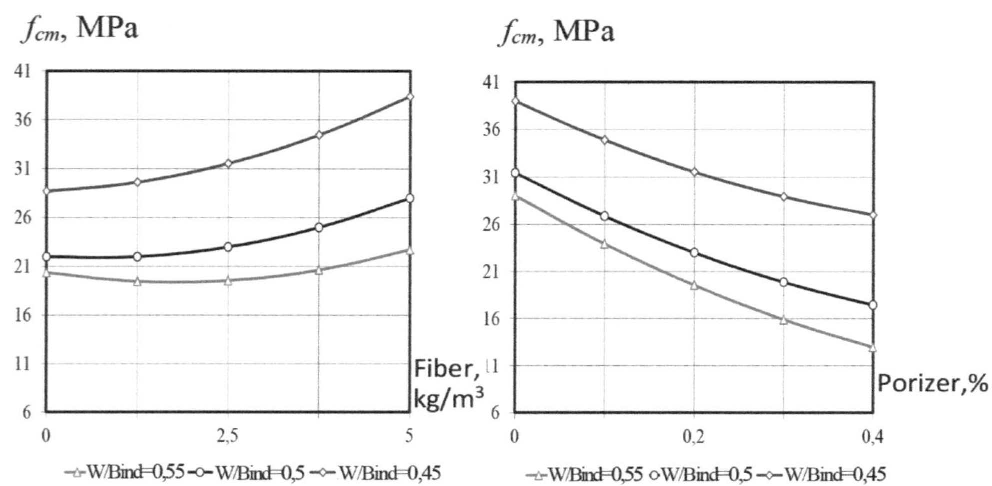

Fig. 4.12 Influence of varied factors on the compressive strength of mortars for 3D concreting made on the GCSIB at the age of 28 days.

References

1. Agustí-Juan, F., Müller, N., Hack, T., Wangler, G., Habert. Potential benefits of digital fabrication for complex structures: Environmental assessment of a robotically fabricated concrete wall. J. Clean. Prod., 154, 330–340. 2017. doi:10.1016/j.jclepro.2017.04.002
2. Ali, M.H., Issayev, G., Shehab, E., Sarfraz, S. A critical review of 3D printing and digital manufacturing in construction engineering. *Rapid Prototyp. J.* 2022. https://doi.org/10.1108/RPJ-07-2021-0160.
3. Anton, A., Reiter, L., Wangler, T., Frangez, V., , Flatt, R.J., Dillenburger, B. A 3D Concrete Printing Prefabrication Platform for Bespoke Columns. *Autom. Constr*., 122. 2021. doi:10.1016/j.autcon.2020.103467.
4. Bentz, D.P., Stutzman, P.E., Zunino, F. Low-Temperature Curing Strength Enhancement in Cement-Based Materials Containing Limestone Powder. *Materials and Structures/Materiaux et Constructions*, *50*. 2017. doi:10.1617/s11527-017-1042-6.
5. Chen, L., He, Y., Yang, Y., Niu, S., Ren, H. The research status and development trend of additive manufacturing technology. *Int. J. Adv. Manuf. Technol.*, *89*, 3651–3660. 2017. https://doi.org/10.1007/s00170-016-9335-4.
6. Chen, Y., Chaves Figueiredo, S., Li, Z., Chang, Z., Jansen, K., Çopuroğlu, O., Schlangen, E. Improving Printability of Limestone-Calcined Clay-Based Cementitious Materials by Using Viscosity-Modifying Admixture. *Cem. Concr. Res.*, 132. 2020. doi:10.1016/j.cemconres.2020.106040.
7. Chen, Y. Li, Z. Figueiredo, S.C. Çopuroğlu, O. Veer, F. Schlangen, E. Limestone and Calcined Clay-Based Sustainable Cementitious Materials for 3D Concrete Printing: A Fundamental Study of Extrudability and Early-Age Strength Development. *Applied Sciences* (Switzerland), *9*. 2017. doi:10.3390/app9091809.
8. Rushing, T.S., Stynoski, P.B., Barna, L.A. Al-Chaar, G.K. Burroughs, J.F. Shannon, J.D., Kreiger, M.A., Case, M.P. Investigation of Concrete Mixtures for Additive Construction. In: *3D Concrete Printing Technology*. Elsevier, pp. 137–160. 2019.
9. S. Bhattacherjee, S., A.S. Basavaraj, A.V. Rahul, M. Santhanam, R. Gettu, B. Panda, E. Schlangen, Y. Chen, O. Copuroglu, G. Ma, L. Wang, M.A. Basit Beigh, V. Mechtcherine. Sustainable materials for 3D concrete printing Cem. *Concr. Compos*., 122 Article 104156. 2021. doi:10.1016/j.cemconcomp.2021.104156.
10. Dvorkin, L., Konkol, J., Marchuk, V., Huts, A. Efficient, Fine-Grained Fly Ash Concrete Based on Metal and Basalt Fibers. Materials, *16*, 3969. 2023. doi:10.3390/ma16113969.

11. Dvorkin, L., Marchuk, V., Hager, I., Maroszek, M. Design of Cement–Slag Concrete Composition for 3D Printing. *Energies* (Basel) 15. 2022. doi:10.3390/en15134610.
12. Dvorkin, L., Marchuk, V., Mróz, K., Maroszek, M., Hager, I. Energy-Efficient Mixtures Suitable for 3D Technologies. *Appl. Sci., 14*, 3038. 2024. doi:10.3390/app14073038.
13. Bolte, G., Zajac, M., Skocek, J., Haha, M.B. Development of composite cements characterized by low environmental footprint. *J. Clean. Prod., 226*, 503–514. 2019. https://doi.org/10.1016/j.jclepro.2019.04.050.
14. Boscaro, F., Quadranti, E., Wangler, T., Mantellato, S., Reiter, L., Flatt, R.J. Eco-Friendly, Set-on-Demand Digital Concrete. *3D Print Addit. Manuf. 9*, 3–11. 2022. doi:10.1089/3dp.2020.0350.
15. Chaves Figueiredo, S., Romero Rodríguez, C., Ahmed, Z.Y., Bos, D.H., Xu, Y., Salet, T.M., Çopuroğlu, O., Schlangen, E., Bos, F.P. An Approach to Develop Printable Strain Hardening Cementitious Composites. *Mater. Des., 2019*, 169. 2019. doi:10.1016/j.matdes.2019.107651.
16. Dvorkin, L., Konkol, J., Marchuk, V., Huts, A. Effectiveness of Polymer Additives in Concrete for 3D Concrete Printing Using Fly Ash. Polymers, 14, 5467. 2022. doi:10.3390/polym14245467.
17. Perrot, A. *3D Printing of Concrete: State of the Art and Challenges of the Digital Construction Revolution* (1st Edn.). Perrot, A. (Ed.)., ISTE Ltd. and John Wiley & Sons Inc.: Hoboken, N.J., USA., 2019, 176 pp.
18. Perrot, A., Rangeard, D., Courteille, E. 3D Printing of Earth-Based Materials: Processing Aspects. *Constr. Build. Mate*r., *172*, 670–676. 2018. doi:10.1016/J.CONBUILDMAT.2018.04.017
19. Khorasani, M., Ghasemi, A., Rolfe, B., Gibson, I. Additive manufacturing: A powerful tool for the aerospace industry. *Rapid Prototyp. J. 28*, 87–100. 2022. https://doi.org/10.1108/RPJ-01-2021-0009.
20. Ma, G., Li, Z. Wang, L. Printable Properties of Cementitious Material Containing Copper Tailings for Extrusion Based 3D Printing. *Constr. Build. Mater.*, 162. 2018. doi:10.1016/j.conbuildmat.2017.12.051.
21. Nerella, V.N., Krause, M., Mechtcherine, V. Direct Printing Test for Buildability of 3D-Printable Concrete Considering Economic Viability. *Autom. Constr.*, 109. 2020. doi:10.1016/j.autcon.2019.102986.
22. Dvorkin, L., Dvorkin, O., Ribakov, Y. *Mathematical Experiments Planning in Concrete Technology*. Nova Science Publishers: New York, NY, USA, 173 pp. 2011.
23. Chen, Y., Veer, F., Çopuroğlu, O. A Critical Review of 3D Concrete Printing as a Low CO2 Concrete Approach. *Heron*, 62. 2017.
24. Sanjayan, J., B. Nematollahi; M. Xia; L. Wang; M. Guowei 3D Concrete Printing with Low Carbon Cements. *Cem. Concr. Res*. 2020.
25. Kaszynska, M., Hoffmann, M., Skibicki, S., Zielinski, A., Technan, M., Olczyk, N., Wroblewski, T. Evaluation of suitability for 3D printing of

high performance concretes. (Vol. 163). 2018. ResearchGate. https://doi.org:10.1051/matecconf/201816301002.
26. Kazemian, A., Yuan, X., Cochran, E., Khoshnevis, B. Cementitious Materials for Construction-Scale 3D Printing: Laboratory Testing of Fresh Printing Mixture. *Constr. Build. Mater.*, 145. 2017. doi:10.1016/j.conbuildmat.2017.04.015.
27. Hela R., Orsakova D. The Mechanical Activation of Fly Ash. *Procedia Engineering*, . *65*, 87– 93. 2013.
28. EN-196.
29. EN-451.
30. Dvorkin L., Dvorkin O., Ribakov Y. *Multi-Parametric Concrete Compositions Design*. Nova Science Publishers, New York, 223 pp. 2013.
31. Hager, I., Maroszek, M., Mróz, K., Kęsek, R., Hebda, M., Dvorkin, L., Marchuk, V. Interlayer Bond Strength Testing in 3D-Printed Mineral Materials for Construction Applications. *Materials*, 15. 2022. doi:10.3390/ma15124112.
32. Bazhenov, Y.M. 1987. *Concrete Technology.* Vysshaya shkola, Moscow. 449 pp. (in Russian).
33. Thomas, M. Optimizing the use of fly ash in concrete. *Portland Cement Association, 2*, 1–24. 2014.
34. Berry, E.E., Malhotra, V.M. 1982. Fly ash for concrete. Critical Review. *ACI Journal, 2*(3), 59–73. 1982.
35. Ondova, M., Stevulova, N., Fecko, L. Fly ash influence on the chemical and mechanical properties of cement concrete cover of pavement. In: *Proceedings of Annual Conference of Materials Research Society of Serba*i, Yucomat, at Montenegro. 164 pp. 2010.
36. Lane. R.O, Best, S.F. Properties and use of fly ash in Portland cement concrete. *Concrete International, 4*(7), 81–92. 1982.
37. Basin, V. 1981. *Adhesive Strength.* Chemistry, Moscow, 208 pp. (in Russian). 1981.
38. Me Bain S.V. *Adhesive Research Committee Reports*. H.M. Stationery Office, London. 1922.
39. Me Laren, A.D. Symposium: *Adhesion and Adhesives*. Sohn Welley and Sons, pp. 57. 1954.
40. Ramachandran, V. S. Adsorption and hydration behavior of tricalcium aluminate-water and tricalcium aluminate – gypsum-water systems in the presence of superplasticizers. J. American Concrete Institute. *80*, 235–241. 1983.
41. De Bruyne, N.A. *Aircraft Eng*., *16*, 115 (1944), *18*, 1 (1947).
42. Sychev M.M. Inorganic adhesives / M.M. Sychev - L.: Chemistry, 1974. - 156 p. 1. Sychev M.M. Inorganic adhesives / M.M. Sychev - L.: Chemistry, 1974. - 156 p. (in Russian).
43. Deryagin B.V., Krotova N.A. Adhesion / B.V. Deryagin, N.A. Krotova.- Ed. USSR Academy of Sciences, 1949.- 12 p. (in Russian).

44. Krotova N.A. About gluing and adhesion / N.A. Krotova.- Ed. USSR Academy of Sciences, 1956.- 91 p. (in Russian).
45. Moskvin N.I. Physico-chemical foundations of gluing and adhesion processes / N.I. Moskvin.- Ed. «Forest Industry», 1964. (in Russian).
46. Kardanov AV Synthetic Adhesives / AV Kardanov - Moscow: Chemistry, 1968. – 584 p. (in Russian).
47. Basin VE Adhesive Strength. - Moscow: Chemistry, 1981. – 208 p. (in Russian).
48. Bouras, R., Mohand, C.S.I.H., Sonebi, M. Adhesion and rheology of joints fresh mortars. *Journal of Materials and Engineering Structures*, *6*(2), 157–165. 2019.
49. *Dvorkin, L., Dvorkin, O.* Basics of Concrete Science. Stroybeton, St. Petersburg, 686 pp. (in Russian). 2006.
50. Dvorkin, L.I., Solomatov, V.I., Vurovoy, V.N., Chudnovskiy, S.M. *Cement Concrete with Mineral Fillers*. Budivelnyk, Kiev, 137 pp. (in Ukrainian). 1991.
51. Xu, Z., Linzhy, L. Research on superfine fly ash and its activity. *Beijing International Symposium on Cement and Concrete*, Beijing, China, May 14–17. *1*, 493–507. 1986.
52. Batrakov, V.G. *Modified Concrete. Theory and Practice*. Moscow, 768 pp. (in Russian). 1998.
53. Dvorkin. L.Y.. Dvorkin. O.L.. Yu, Garnitsky, V, Rizhenko, I.M. *Modified Ash-Containing Dry Mortars for Masonry and Adhesive Mortars*. Monograph. Rivne, 219 pp. (in Ukraine). 2013.
54. Dvorkin, L., Dvorkin, O. *Basics of Concrete Science: Optimum Design of Concrete Mixtures*. Amazon. (Kindle edition)/(e-book), Stroi-Beton, S-Peterburg. 382 pp. (in Russian). 2006.
55. Ramachandran, V.S., Feldman, R.F., Kollepardi, M. et al. *Additives in Concrete: Reference Book*. Moscow. Stroyizdat. 575 pp. (in Russian). 1988.
56. Neville, A.M. Properties of Concrete (4th Edn.) John Wiley & Sons, New York, 844 pp. 1996.
57. Dvorkin, L., Gotz, V., Dvorkin, O. *Testing of Concrete and Mortar. Design Composition*, 397 pp. (in Ukrainian). 2014.
58. Stolnikov, V.V. Use of fly ash from the combustion of pulverized fuel in thermal power plants. 50 pp. (in Russian). 1989.
59. Gorchakov, G.I., M.M. Kapkin, B.G. Skramtaev. *Increasing the Frost Resistance of Concrete in the Structures of Industrial and Hydraulic Structures*. Moskow: Stroizdat. 195. pp. 1965.
60. Petrov BA, Sidyakov PV Dedusting of Process Gases in Cement Production. Leningrad – Moscow: Stroyizdat – 1965. – 90 p. (in Russian).
61. Kolenova KG Study of the Composition and Properties of Dust Captured by Rotary Kiln Electrostatic Precipitators and Ways of Using It. Abstract of Cand. Sci. (Medicine) Dissertation. Moscow: - 1962 – 25 p. (in Russian).

62. Banit FG, Malygin AD Dust Collection and Gas Cleaning in the Building Materials Industry. – M.: Stroyizdat – 1979 – 351 p. (in Russian).
63. Gus S., Keil F. Behavior of alkalis during cement firing. Coll. Cement. – M.: Publishing House of Foreign Literature, 1962, pp. 37-69. (in Russian).
64. Henning O., Stieler R. Use of highly alkaline cement dust in Portland cement// Proceedings of the 6th International Congress on Cement Chemistry. G.2. M.: Stroyizdat, 1974 – pp. 27-30. (in Russian).
65. Butt Yu.M., Sychev M.M., Timashev V.V. Chemical technology of binders. – M.: Higher School, 1980. – 472 p. (in Russian).
66. Forinton. J. Recycling kiln bypass dust into valuable materials. In: *IEEE-IAS/PCA, Cement Industry Technical Conference,* Orlando, 240 pp. 2013.
67. Ternovaya E.A. Use of Cement Plant Rotary Kiln Dust// Construction Materials, Products, and Sanitary Equipment. 1987, No. 10, pp. 22–23. (in Russian).
68. Spesivtsev, Yu. A. Production of a Local Binder Based on Rotary Kiln Electrostatic Precipitator Dust// Construction Materials. 1963, pp. 18–19. (in Russian).
69. Batalin, B. S. Local Binder Based on Novo-Pashysky Cement Plant Electrostatic Precipitator Dust. Collection of Binding Materials of Siberia and the Far East. Novosibirsk: Nauka, 1970, pp. 283–287. (in Russian).
70. Glukhovsky, V. D., Pakhomov, V. A. Slag-Alkali Cements and Concretes. K.: Budivelnik, 1978, 174 p. (in Russian).
71. Glukhovsky V.D., Makedon N.L. Soil cements containing fly ash from clinker kilns// Slag-alkali cements, concretes and structures. Abstracts of reports. Kyiv, 1989 – pp. 35-36. (in Russian).
72. Dvorkin, L.I., O.L. Dorkin. *Building Mineral Binders.* Moskow: Infra-Engineering. 544 pp. 2011.
73. Powers T.C., Brownyard T.L. Studies and physical properties of hardened Portland cement paste. *Proc. Amer. Concrete Inst.*, №43. 1947.
74. Bouras, R., Mohand, C.S.I.H. Sonebi, M. Adhesion and rheology of joints fresh mortars. *Journal of Materials and Engineering Structures, 6*(2), 157–165. 2019.
75. Collepardi, M., Corradi, M., Valente, M. Influence of polymerisation of sulfonated naphtaline condensate of the interaction with cement: Developments in the use of superplasticizers. *ACI,* 485–498. 1981.
76. Costa, U., Massazza, F, Barrila, A. Adsorption of superplasticizers on C3S, changes in zero potential, and rheology of pastes. *Cemento, 79*(4), 323–336. 1982.
77. Batrakov, V.G. 1990. *Modified Concretes.* Stroyizdat, Moscow, 396 pp. (in Russian). 1990.
78. Batrakov, V.G.. Babaev, S.G., Bashlykov, N.F. Concretes on binders with low water demand. *Concrete and Reinforced Concrete,* (11), 4–6. 1998.

Index

For Product Safety Concerns and Information please contact our EU representative GPSR@taylorandfrancis.com Taylor & Francis Verlag GmbH, Kaufingerstraße 24, 80331 München, Germany